W9-AMB-581

How to Pick a Peach

Russ Parsons

How to Pick
a Peach

THE SEARCH FOR FLAVOR
FROM FARM TO TABLE

Houghton Mifflin Company • BOSTON • NEW YORK • 2007

Copyright © 2007 by Russ Parsons

All rights reserved

For information about permission to reproduce
selections from this book, write to Permissions,
Houghton Mifflin Company, 215 Park Avenue South,
New York, New York 10003.

Visit our Web site: www.houghtonmifflinbooks.com.

Library of Congress Cataloging-in-Publication Data
Parsons, Russ.
   How to pick a peach : the search for flavor from farm
to table / Russ Parsons.
        p.   cm.
   Includes index.
   ISBN-13: 978-0-618-46348-0
   ISBN-10: 0-618-46348-8
   1. Vegetables. 2. Fruit. 3. Cookery (Vegetables)
4. Cookery (Fruit) I. Title.
   TX391.P37 2007
   641.3'5 — dc22   2006035462

Book design by Melissa Lotfy

Printed in the United States of America

MP 10 9 8 7 6 5 4 3 2 1

Refrigeration list on page 205 adapted from *Postharvest
Technology of Horticultural Crops,* edited by Adel Kader,
copyright 2002 by the Regents of the University of California, Division of Agriculture on Natural Resources.

For all the talented farmers
who work hard
so we cooks don't have to

Acknowledgments

A book like this is the product of so many minds that it's hard to know where to start in thanking them. Legions of people were incredibly generous with their time and knowledge, and any errors are undoubtedly the result of my misunderstanding and no fault of theirs. In this case, *do* blame the messenger.

It's only fitting to start with the farmers: Laura Avery, Joe Avitua, Ed Beckman, Maryann and Paul Carpenter, Jim Churchill, Bill Coleman, Vance Corum, the Gean-Iwamoto family, Lucio Gomiero, Jim Howard, Fitz Kelly, Art Lange, Marc Marchini, Mas Masumoto, Richard Matoian, Phil McGrath, Jon Rowley, the Tamai family, Tony Thacher, the Weiser family, the Zuckerman family.

And then, of course, the cooks: Gino Angelini, Michael Cimarusti, Josiah Citrin, Jim Dodge, Vincent Farenga and Nicole Dufresne, Mark Furstenburg, Alain Giraud, Suzanne Goin, John and Leslie House, Thomas Keller, Evan Kleiman, Eric Klein, David LeFevre, Deborah Madison, Drake McCarthy and Jeanne Laber, Mark Peel, James Peterson, Michel Richard, Michael Roberts, Judy Rodgers, Sonoko Sakai, Fred Seidman and Sherry Virbila, Lindsey Shere, Ken Shoemaker and Trudy Baker, Martha Rose Shulman, Nancy Silverton, Maria and Rob Sinskey, David Tanis, Paula Wolfert, Tim Woods, Clifford Wright, Pauline and Luciano Zamboni.

I've learned something from almost every writer I've ever read, but especially: Toni Allegra, Julia Child, Amy Goldman, Emily Green, Amanda Hesser, David Karp, Matt Kramer, Patric Kuh, Mike Madison, Harold McGee, Charles Perry, Ruth Reichl, Phyllis Richman, Michael Ruhlman, Elizabeth Schneider, David Shaw, Rod Smith, Jeffrey Steingarten, Steven Stoll, Sylvia Thompson, Anne Tyler, Richard Walker, Faith Willinger. And finally, the ever argumentative folks at eGullet, who never fail to give me something to think about.

There is a tremendous amount of academic work available on fruits, vegetables and agriculture in general, if you know where to look. I'd especially like to thank: all the unsung workers at the Department of Agriculture's Economic Research Service and National Agricultural Statistics Service, the University of California at Davis, Marita Cantwell, Carlos Crisosto, Kevin Day, Stephen Facciola, Julie Guthman, Adel Kader, Kirk Larson, Margaret McWilliams, Harry Paris, Mikeal Roose, Richard Lance Walheim, Michael Yang.

I've been terrifically fortunate in my professional life to have worked with some of the best people around. At the *Los Angeles Times*: John Carroll and Dean Baquet, who took a dispirited newspaper and turned it into one of the best in the country; John Montorio and Michalene Busico, who gave me the best job in food journalism; Leslie Brenner for years of good editing; Donna Deane, Mayi Brady and Maryellen Driscoll for keeping the recipes straight; and all of the rest of the food staff, past and present.

I've been just as fortunate in the book business. *How to Read a French Fry* was a dream come true for me — all I wanted to do was write a good book, and I hoped it would sell well enough so my friend and editor Rux Martin wouldn't get fired. It did better than I'd ever dreamed, and I'd like to thank Rux (still working; still keeping the Post-it company in business), my agent Judith Weber, the publicist Deb DeLosa, the copyeditor Barb Jatkola and everybody else at Houghton Mifflin for making the impossible happen.

Finally, as always, for my "girlies": Kathy and Sarah.

Contents

Fall

· · · · ·

Winter

· · · · · · · ·

The Vegetables and Fruits Alphabetically

The Recipes by Category

SALADS

MAIN DISHES

How to Pick a Peach

Introduction

If, as nineteenth-century French chef Antonin Carême so famously remarked, the cuisine of his day should have been most rightly regarded as a branch of architecture, today's culinary fashion would be much more likely to send every aspiring chef scurrying straight for the nearest college of agriculture. There has been a revolution in American kitchens as a whole generation of cooks has discovered the startling idea that a great dish can only be built on good ingredients. On the one hand, this is welcome, in fact overdue. For too long we have ignored where our food comes from and what it takes to grow it.

Too often, though, in our conversations, the complex issues of agriculture have been reduced to easy catchphrases: organics, crop subsidies, corporate farming, genetically modified organisms, agricultural-industrial complex — often by people who have never visited a farm (or even planted a backyard garden). These are repeated like mantras until, like all mantras, the actual words lose their meaning.

"Eat local; eat seasonal." How many times have you heard it? Certainly, it seems like an idea that should have been obvious to us from the start. After all, it is exactly what all good cooks have done since time immemorial. If it now seems new, it is only because we have spent the last couple of generations trying to forget it. This wasn't so much a conscious effort as a slow accretion of bad habits encouraged by modernization. Comfort, ease, efficiency and economy are wonderful things in most cases, but their effect on cooking has been devastating.

At its heart, cooking is a primitive act and remarkably simple. You choose what seems tastiest, and then you try to make it better. Seems easy, right? Only until you try to do it yourself. Let's say you want to make a peach pie. You go to the store and buy some fruit. You pick the brightest, reddest ones and stick them in your cart. You take them home, peel them and slice them, and then sweeten them with a little sugar and dust them with a little spice. And then you taste them. Nothing. So you add more sugar and you add more spice. And then you do that again. And still it doesn't turn out the way you want. By the time you pull the pie from the oven, you might as well call it sugar-and-spice pie, because you sure can't taste any peaches. So what happened? There are many possible culprits, stretching from the tree the fruit was grown on to your home kitchen.

This book will examine some of the possibilities. It will help you sort through the fruits and vegetables you find in the market (both super and farmers'), telling you how to select, store and prepare them. And it will shine a little light on some contentious issues by taking you right out to the fields and introducing you to the people involved.

The first thing you have to understand is that the whole idea of eating locally and seasonally is not based merely on some philosophical framework. It may indeed be good for the planet, but that is for greater minds to decide — I'm mainly interested in fixing a good dinner. And believe it or not, eating locally and seasonally is backed up by some very basic scientific principles.

Fruits and vegetables are not manufactured items that remain the same throughout their shelf life. They age and change just like the rest of us. Some of them even improve. Fruits such as peaches, tomatoes and some melons will actually finish ripening after they've been harvested — as long as you treat them right. Most fruits and vegetables, though, do not improve.

From the moment they are picked, they go into decline, long and slow in some cases and dismayingly rapid in others. Lettuces and herbs, for example, are really made of nothing but extremely thin sheets of plant material plumped up by water. As soon as they are picked, the water begins to evaporate and the leaves begin to wilt. To control that loss of moisture, we keep them cold (moisture loss happens faster with warmth), and we keep them in tightly confined spaces (these become humid quickly, discouraging the leaf from giving up more moisture).

Greens are just an extreme example of what happens to every fruit and vegetable eventually, even the hardiest ones. How many times have you found a carrot, lost behind some containers on the bottom shelf of the refrigerator, that has all the crispness of an overcooked spaghetti noodle?

The farther away from you something is grown, the longer it must spend in some truck or railroad car getting to you, and the greater are the odds that what eventually arrives will be less than what it could have been. The agriculture business has gotten very good at keeping food as fresh as possible along the way, but there is no arguing with time.

As growers and marketers try to come up with ways to beat the aging process, one favorite technique is picking fruit earlier. To understand why they do this, let's look at the one surefire way to pick a perfect peach. All you have to do is go out into your backyard, find the one on the tree that is at the most perfect point of maturity and ripeness, pluck it gently from its branch, cradle it gingerly in your cupped hands and then walk quickly to your kitchen. Don't run — you might jostle it. A peach like this is a treasure, a taste to remember all your life. Of course, this method works only if you live someplace with the right combination of soil and climate to grow great peaches — and if you happen to be a great gardener.

For reasons of time, inclination, talent or real estate, that is impractical for most of us. So we start to make compromises. The most basic one is delegating the growing of our peaches to someone someplace else — we don't really know who or where. This is not bad. In fact, it is one of the great blessings of living in the modern age — our forefathers worked for generations so we wouldn't have to labor in the fields. But at the same time, we have to recognize that in making these compromises, we have made sacrifices. And for every step that we add between ourselves and the soil, we move one step farther from that ideal peach and introduce one more opportunity for misadventure.

One thing that has not changed after all this progress is the essential act of growing. Plants work the same way whether they're planted on your back patio or on four hundred acres outside Fresno. And even though we may have forgotten, growing food is not the same as manufacturing widgets. It is an endeavor that requires the right circumstances, talent and more than a little bit of luck. It is also an endeavor fraught with the potential for mishap. Bad seed, bad soil, bad weather — all are waiting in line to surprise the unlucky farmer. And then even after the fruit has cleared those hurdles, there's always the chance that some lunkhead will pop a thumb through that lovely little peach in the warehouse.

We don't usually think about these things, good or bad, when we go to the grocery store — they happen without our ever knowing about it. In the produce section, peaches simply appear as if by magic. But there is no magic. Good food comes through talent and hard work. And when something goes wrong, someone has to take responsibility.

That is usually the farmer. Any bad luck that befalls a crop is all his. So to protect himself and reduce his risk as much as possible, he, too, begins to make compromises. Maybe he chooses to grow types of peaches that are more resistant to disease or that produce more fruit — even if they don't have the best flavor. Maybe he shades his farming practices toward growing more peaches rather than better ones. In commercial agriculture, there is precious little financial reward for flavor, but you do get paid by the pound (or, more accurately, the case). Maybe he picks the peaches a little early. Fruit that is in the warehouse never gets ruined by rain or hail, and if it's a little firm, that only means it will ship better.

And so we may wind up with that flavorless peach. But who among us can blame the farmer for making those choices? Farming is, after all, a business, and one with a perilously slim profit margin. Forget what you pay for food at the grocery store. On average growers earn only about 20 percent of the retail price of what they harvest. And those farm subsidies you've heard so much about — the ones that guarantee farmers a certain price for their crops or even pay them not to plant? Those apply only to crops that are used for manufacturing — cotton, sugar, field corn and the like. There are no direct subsidies for fruit and vegetable farmers.

It's no wonder that most people don't realize these things. They're almost encouraged not to. The produce department at a modern supermarket — the only place most of us come into contact with farming — is by design a place outside of nature, unbound from the laws of climate, location and seasonality. It is a place where you can find strawberries and grapes in December and apples and cabbages in July. It is a place where you can buy delicate greens in Phoenix in the summer and tropical fruits in Minneapolis in the winter. Indeed, it is a place where you would have trouble telling which of those cities you were in and what time of year it was without stepping outside.

The average shopper in an American supermarket enjoys a bounty that was beyond the dreams of even the richest and most powerful people only fifty years ago. Pineapples jetted in from Hawaii? No problem. Peaches and plums delivered from Chile in January? You bet. Tuscan black kale? It's on sale. Literally. American agriculture is unsurpassed when it comes to delivering a lot of food at a very low price. The average American today spends less than 11 percent of his or her disposable income on food, less than in any other industrialized nation and less than half what our grandparents spent before World War II. (Still, a recent poll found that for an overwhelming majority of Americans, their biggest concern about food is that it costs too much.)

But at the same time, we cooks are stuck with the result, which frequently means that we're forced to choose among fruits and vegetables that don't taste like anything. What are we to do? How do we pick the best of what is available? The answer is both simple and complicated. Boiled down to its essentials, choosing the best produce means selecting that which looks most like it came out of your garden. Usually — though not always — this means picking what was grown closest to you, in terms of both location and season.

This is the kind of thing that drives some people nuts. Those who live in areas that are, shall we say, agriculturally or climatically challenged point out quite correctly that this would mean eating parsnips and potatoes for half the year. But choosing the best local and seasonal produce is a goal, not a dogma. As long as you understand the reasoning behind it, you can make an informed decision, and what you choose to do is between you and your conscience — and the palates of the people you're feeding.

Granted, this whole local and seasonal thing is something that comes more easily and naturally in California. In fact, there is a very good reason it began here. When I first moved to the state more than twenty years ago and began writing about farming from a cook's-eye view, one of the first things that struck me was the intimate connection between the produce I found in the supermarket and the local weather. At first it came as a shock to realize that when we enjoyed one of those rainy spring weekends, it was hard to find good strawberries for the next few days. Like most Americans, I had never lived anyplace where much of the food in the store was actually grown nearby. In California it's difficult to avoid that connection.

Although California's "local/seasonal" orientation may be leading the way to better eating, there is no denying that the state is also a major part of the problem in the first place, because this is where much of the substandard produce in the supermarket is grown. Just as Hollywood dominates the film industry, so the rest of the state rules commercial farming. By itself, California produces more than half of all the fruits and vegetables grown in this country. The second-place state, Florida, grows less than 10 percent. California grows more than 70 percent of the national harvest of grapes, garlic, lettuce, strawberries, broccoli, carrots, lemons, plums, celery and cauliflower. For all intents and purposes, it is the sole commercial producer of a dozen crops, including apricots, dates, artichokes, avocados, nectarines and figs.

This dominance is not the result of some vast conspiracy, but a natural result of modern history. As any agricultural historian will tell you, the recent return to the idea of produce being supplied by local farmers is nothing new. Until the early twentieth century, this was the way all farming was done. Cities were densely packed and intensely urban. Each was surrounded by greenbelts of farmland devoted to growing the food that was needed to keep the local folks going. In those days eating locally and seasonally was a necessity, not a lifestyle choice. The great majority of what you ate had to be grown within a one-day carriage ride of your home. Furthermore, the person you bought the produce from was usually the neighborhood grocer — almost always an independent operator who owned his own store and contracted with local farmers for his stock.

Three separate but intimately intertwined developments changed farming in America, resulting in the great agricultural upheaval of the twentieth century. The first of these was a revolution in transportation, including the establishment of the cross-country railroad and the invention of the refrigerated railcar, as well as the building of local light rail for commuters. The first two allowed produce to be shipped from much farther away than a city's immediate surroundings. Although we now consider this to be food's fall from grace, at the time it was regarded as one of the best things that had ever happened.

When the first strawberries were successfully shipped from California to the East in the 1890s, banquets were held in every major city along the way to celebrate the event. Considering the likely condition of those berries — given the speed of trains in those days and the fact that they were refrigerated by means of ice piled into compartments on top of the cars (the trains would have to stop several times a day to get more ice) — we might wonder at the ecstatic accounts of this great advance. But then, of course, we aren't forced to subsist on dried apples and cabbage for several months of the year as people back then were. The establishment of the national highway system after World War II sealed the deal.

As transportation brought the nation, and its fruits and vegetables, closer together, it also allowed the cities to disperse. Light rail lines and highways freed people from having to live within walking distance of their work. Thousands of souls, eager to break free of crowded city centers that were then decried as being rife with crime and disease, rode these commuter lines to rapidly developing suburbs. This was land that until that time had been fit only for farming. But then, as now, the price of real estate had almost as much to do with agricultural economics as the price of produce. Farmers, beginning a century-long pattern that continues to this day, took the sure money and sold off their land to developers, either quitting agriculture altogether and finding another — hopefully easier — line of work, or moving progressively farther outside the city, frequently only to sell out and relocate again.

At the same time that the country was being brought closer together — at least produce-wise — and farmers were getting chased farther from their original locations, the great agricultural fields of California's Central Valley were being planted. Originally, this area was divided into huge tracts of land given as grants by the Spanish kings to those they believed were deserving of favor. Because little water is available for much of the year, the land had traditionally been used for running cattle. Cities — even towns — were few and far between.

But the twentieth century changed all that. New technology allowed irrigation to bring water to much of the land. Between the original families dividing and selling off portions of their land over time and ambitious investors from the East buying up acreage for speculation, what had once been endless acres of rangeland was transformed into family farm plots.

The railroad had a hand in this, too, particularly after buying up large amounts of land on the valley's eastern side, where water was more available. This land was split up into even smaller portions and sold off at low prices to encourage the establishment of farms — farms that would need to ship their goods by rail. The railroads even established market towns along the main route where farmers could buy the supplies they needed — also shipped in by rail. To this day towns down the spine of the Central Valley are spaced evenly apart — just far enough for what once would have been one day's journey.

What the Central Valley might once have lacked in water, it more than made up for in weather. California has one of the few examples of a true Mediterranean climate in North America. Temperatures are mild, and rains come almost entirely in the winter offseason. Summers are warm and dry — perfect for ripening fruits and vegetables with a minimum threat of disease. Even better, because winter temperatures are so moderate — January temperatures average 54 degrees — farmers can frequently get two, three or even four crops from the same piece of ground in a single year. Between the perfect weather and the huge expanses of land available for farming, the Central Valley was an agricultural paradise just waiting for the planting.

And plant farmers did. Between 1895 and 1945 California's total fruit production increased almost 1,000 percent. Today the Central Valley alone, a 1,400-square-mile strip spread over just eight counties, accounts for as much as a quarter of the total agricultural production of the United States. Four of the top five agricultural counties in the country — Fresno, Tulare, Kern and Merced — are located in the valley, and the fifth — Monterey — is right next door.

Complementing all of this plentiful produce and the improved means of delivery, the grocery business went through its first wave of consolidations as the biggest local proprietors bought out their less successful competitors and expanded into their competitors' stores. The great innovation here was the self-serve grocery. Up until that time, the fruits and vegetables in a store would be stacked behind the counter, to be fetched by the proprietor or an employee when the customer requested them.

Clarence Saunders changed all that when in 1916 he opened his first Piggly Wiggly store in Memphis and allowed customers to wander the aisles, selecting for themselves the things they wanted to buy. This not only proved to be cheaper than the previous method, but it also allowed more customers to be served in a given amount of time. The idea seems utterly commonplace today. (In fact, the idea of a store where you have to ask for someone behind a counter to give you a carrot, as if it were a prescription drug, seems ludicrous.) But at the time, it was so revolutionary that Saunders was able to patent it.

Another grocer, Michael Cullen, took marketing to the next level in 1930 by opening his first King Cullen store in Queens, New York, a true supermarket that measured 6,000 square feet — gargantuan for that time. (In 2005 the median size of a supermarket was nearly 50,000 square feet.) Cullen also took cost cutting to the next level. His motto — "Pile it high. Sell it low" — summed up the new approach with remarkable brevity. By 1936 Cullen owned seventeen stores.

The next year the venerable Great Atlantic & Pacific Tea Company, a loose confederation of groceries founded in 1859, began making the first moves toward a real supermarket chain, converting many of its 15,000 smaller stores into A&P supermarkets. By the end of the decade, there would be 16,000 of them, and they would gross $1 billion.

A grocery chain operates very differently from an independent store — particularly when the ultimate goal is having the lowest prices. First of all, instead of having direct deliveries to every store, there needs to be a central clearinghouse for the produce to flow through. This adds a day or two to the transportation. Because supplying even a modest chain takes much more than any single farm — or even group of farms — can grow, the chain has to rely on middlemen at the supply end who can gather the produce of enough farms to ensure that the flow of fruits and vegetables is uninterrupted and adequate to serve several million customers. This introduces another level of bureaucracy to the produce distribution system. It means one more group making a profit, putting even more downward pressure on the prices paid to farmers, and it adds another day or more to the time it takes for fruits and vegetables to get from farm to table.

Maybe even more to the point, it introduces another level of risk-averse decision makers. Because when you think about it, farmers don't really sell their fruits and vegetables to the people who eat them. They sell them to the people who sell them to the people who eat them (and often there are several more levels in between). Although great flavor is the quality we eaters hold most dear in our food, for everyone else involved what matters is profitability. And that almost inevitably means trying to achieve longer shelf life. When you're operating on razor-thin margins, it's painful to throw away food you've already paid for because it has gone over the hill.

Let's use the publishing industry as a metaphor. If authors were pressured to write books more quickly so those books could sit on the store shelves for a longer time during the season in which they were published, you can see that compromises would begin to be made. "Mr. Faulkner, if only your prose weren't so perfectly ripe, we could pay you 20 percent more." *Absalom, Absalom!* Might have ended up as *Great Abs!*

When you combine a seemingly insatiable hunger for low-priced food, a vast fertile growing area ready to be tapped and the transportation network that links them all together, it seems inevitable that you wind up with something like the state of modern agriculture. As the focus on farming moved more and more away from the old decentralized model to the new West Coast version, folks left the farms in droves. In 1900 more than 60 percent of Americans lived in rural areas. By 1920 that number had dropped to 48 percent, by 1940 to 43 percent and by 1960 to 30 percent. Finally, by 2000 less than 20 percent of Americans lived in rural areas.

But that's only part of the story. Those people who did stay on the farm saw their incomes plummet at the same time the prices of their farmland were skyrocketing. Between 1947 and 1987 agricultural land prices increased 75 percent, while real commodity prices declined 60 percent. To get out of this squeeze, farmers tried different things. Some of them got bigger, hoping to take advantage of economies of scale. (Between 1959 and 1969 the average size of an American farm increased 77 percent.) Technological advances in farming helped farmers produce more, as new techniques (including chemical pesticides, fertilizers and herbicides) allowed bigger harvests.

Many farmers also took second jobs. The percentage of farmers' household income that came from jobs outside the farm increased from 30 percent in 1950 to more than 75 percent in 1999, as one or both partners took on outside work. And some farmers began looking for ways to get more money for the crops they grew. One way to do that was to move to higher-value crops. Instead of growing cotton or cattle, for example, they'd grow peaches or strawberries, or they'd find a more exotic niche, such as fresh herbs or heirloom tomatoes.

Another way to earn more for their harvests was to find ways they could take home more of the selling price of their produce. This usually meant selling directly to customers rather than going through a middleman. At first farmers did this by setting up roadside stands. This brought in much-needed cash, but sitting by the side of a rural road waiting for someone to drive by was a pretty inefficient form of marketing.

So the next step was finding a central point where farmers could gather and sell their crops. This idea was not new — most towns had had farmers' markets into the 1930s and even later. But selling produce had become hard to do, at least legally. Throughout the second half of the twentieth century, as fruit and vegetable farming became more institutionalized, more regulations were introduced specifying how produce must be packed, transported and sold. These regulations made it more difficult for farmers to simply load up their trucks and head to market. As a result, with only a few exceptions, farmers' markets gradually faded from the scene.

In the 1970s, however, things began to change, thanks to a combination of a dwindling farm economy, a growing back-to-the-earth movement and counterculture marketing. In California and other places, it worked like this: food co-ops popped up in college towns, offering members a chance to beat conventional marketing by pooling resources and buying directly from wholesalers and growers rather than going through the supermarkets.

At the same time, a new breed of farmers — frequently growing unusual crops such as herbs or lettuces in quantities that would normally be too small for commercial sale — began peddling their produce to the co-ops. They found willing buyers. Frequently they were growing fruits and vegetables that their customers couldn't get anywhere else, things that were considered exotic and therefore worthy of a higher price. Farmers may be conservative by nature, but there are always a few rugged individuals who are not afraid to try something new — particularly if there is more money to be made. And thus farmers were reintroduced to the rewards of direct marketing.

There was no single motive behind the resurgence of farmers' markets in the mid-1970s. In New York, the greenmarket movement started as a way to save pressured family farmers and, not incidentally, to revivify some troubled Manhattan real estate (the Union Square market was one of the most important factors in that once scruffy neighborhood's rebirth). In Northern California, the motives stemmed more from an in-your-face countercultural desire to circumvent the established food chain. And in Southern California, the first five farmers' markets were organized by a religious charity to provide high-quality food to the poor. Today, whether by design or not, every market combines all of those goals in varying proportions.

In California, establishing farmers' markets literally took an act of government. State law, designed to smooth interstate agricultural commerce, required growers to pack their fruits and vegetables in certain ways, right down to specifying the kinds of boxes to be used. Furthermore, it forbade them from selling to anyone other than a licensed broker. In 1977, the year after the first New York greenmarket was founded, the California Department of Food and Agriculture, at the urging of then-governor Jerry Brown, introduced new regulations that would exempt farmers from these rules if they could prove that their fruits and vegetables were going to be sold directly to consumers through certified farmers' markets. The original seven California farmers' markets were located exactly where you would expect them to be — in the Bay Area and in college towns near agricultural areas: San Jose, Redding, Davis, Sacramento, Riverside, Santa Rosa and Santa Cruz. The idea caught on quickly, and only two years later there were twenty markets. Two years after that there were more than fifty. And so it went. Today there are more than 350 certified farmers' markets in the state — more than eighty-five in Southern California alone. Many communities have not just one but several, held on different days of the week and in different locations, each with a slightly different mix of farmers.

The phenomenon is not by any means limited to California. Farmers' markets have sprung up everywhere. From 1994 to 2000 the number increased almost 80 percent nationally, to 3,100. From 1987 to 2000 the amount of produce sold at them more than doubled, and from 1994 to 2000 the number of people shopping at farmers' markets nearly tripled, to more than 2.7 million a year.

Granted, farmers' market sales still make up a tiny percentage of the fruits and vegetables bought in this country — probably only a little more than 2 percent. But the impact of the markets extends well beyond direct sales. Consider this: During the 1990s, the period of greatest growth of farmers' markets, American spending on fresh fruits and vegetables more than doubled. The average number of items carried by a supermarket produce department nearly doubled, too, going from 173 items in 1987 to 335 in 1997.

Of course, not all of this increase is because of the influence of farmers' markets, but they do serve as points of introduction for many new items. Things such as heirloom tomatoes, specialty potatoes, salad mixes and fresh herbs were introduced at smaller markets, then adopted by the big boys. This should come as no surprise, since more than 70 percent of the farmers who sell at farmers' markets also sell through regular channels, some of them even leveraging their farmers' market credentials into higher wholesale prices. And fully 20 percent of farmers' markets report having wholesalers shopping there on a regular basis.

Another niche product that started at farmers' markets and has moved to the mainstream is organic produce. In fact, many shoppers believe that everything sold at farmers' markets is raised according to organic standards. That is certainly not true, but it does offer a telling insight. The word "organic" has become for many people a kind of shorthand summing-up of a much more complicated set of attributes. These are probably best described by the scholar Julie Guthman's phrase "agrarian dreams." Often when people say "organic," they think they are describing small farms that are family owned and carefully maintained, following a full complex of enlightened agricultural principles, including everything from crop rotation and the encouragement of beneficial insects to benevolent labor practices and abstention from any chemical pesticides and fertilizers. The reality is that the official definition of the word "organic" — the legal rules that must be followed before a producer can apply that label to his crops — doesn't require any of those things but the last.

First, organic farms can be every bit as big as conventional ones. The organic salad king Earthbound Farm, for example, harvests a total of more than 24,000 acres (a combination of land that it owns and some that it farms in partnerships). Furthermore, according to an analysis by Guthman, conventional fruit and vegetable farms have no higher rate of corporate ownership than their organic peers. Contrary to what seems to be the common impression, roughly 90 percent of all farms in California are owned either by individuals or families, not by large corporations. As for the enlightened agricultural principles, about the only thing the law does is limit the kinds of chemicals that can be used to those that are derived by "organic" means.

This is not an inconsiderable thing, depending on how you assess the risks involved in the use of pesticides, herbicides and fertilizers. But it should be pointed out that in the more than ten years that the Environmental Protection Agency has been testing fruits and vegetables, less than 1 percent of all samples have turned up with chemical residues that violated the established standards. (Some scoffers like to point out that chemical residues also turn up on organically grown produce, but this happens much less frequently and at such low concentrations that it doesn't really seem a fair criticism.)

It's also important to remember that the world of agriculture does not divide cleanly into black-and-white stereotypes of "pure organic" and "chemical junkie." One of the larger benefits of the organic movement has been to demonstrate to conventional farmers the efficacy of many techniques they might not otherwise have tried. Today a huge number of farms fall into a category between organic and conventional, one that is called variously "sustainable" and "natural" and that incorporates many of the precepts of organic farming. These practices include things such as the use of soil-enriching cover crops, managing the weeds between the crop rows to encourage beneficial insects and even setting up birdhouses and bat houses in the fields to help keep bugs down. These farmers do reserve the right to spray when they believe it is necessary and as a result cannot call themselves organic.

Whether this gray area of farming is good enough for you is a personal philosophical choice. As for me, I prefer to take a fairly liberal view. I remember a farmer once pointing out to me that even though we have a tremendous problem with the overprescription of antibiotics, most people aren't becoming Christian Scientists over it. It seems to me that you can pursue a moderate approach without having a guilty conscience. Some of the best farmers I know are organic, but more are not. And when they're being candid, many of the ones who are organic will admit that their main motivation is hoping to be able to earn a little more money.

After years of going back and forth on the issue, I've finally decided simply to trust taste. This is only partly hedonistic. After all, a lazy reliance on the overuse of chemicals can give you mountains of perfect-looking fruit, but it can't give you flavor. That can come only from careful farming.

If you decide that going organic is important to you, you should not begrudge the farmers charging more for their produce. Different crops adapt to organic practices in different ways. Grapes and potatoes, for example, are relatively easy to grow organically. But other things, such as peaches and strawberries, are devilishly difficult, and a farmer must be willing to lose a certain portion of his crops to disease, predation or physical defects. Estimates of exactly how much the farmer will give up vary but generally fall into the 20 to 30 percent range. This obviously means that a farmer will have to be able to realize a 20 to 30 percent premium for his product just to stay even with his competitors who are farming conventionally. Thus far — no matter what you may be paying at retail — organic farmers generally have had to settle for less than that at wholesale and in some cases have had to sell at a discount. A recent survey found that even at farmers' markets, organic growers are having a hard time getting anything extra.

In the end, though, everything comes down to the simple fact that good cooking starts with good shopping. That, of course, is exactly what most shoppers find so frustrating. One of the ironies of modern life is that although we have more ingredients to choose from than ever before, much of it is not worth buying. In a way, you could say that we have been blinded by plenty. We have so many choices that we've forgotten how to select only that which is best. For even fairly sophisticated cooks, the produce section of the neighborhood grocery store is a place of mystery. That there is such lack of information about fruits and vegetables has always seemed to me to be deeply ironic.

Put simply, never has a civilization had so much easy access to so great a variety of foodstuffs at so low a price. But despite that rich variety, the quality of even the simplest things is often poor: tomatoes that taste like cotton; peaches that will never drip; strawberries that could bend a fork. This is the stuff most people try to make dinner from, and that's why cooking frustrates them so.

But it doesn't have to be that way. Armed with some basic information about what to look for when you're shopping for fruits and vegetables, how to store them properly when you get home and a few basics of preparation, you can cook well even with supermarket produce.

Let's go back to that troubled peach. You know how it goes: You buy a peach at the market and take it home, the whole time anticipating the moment when you lean over the sink and bite into it. But when that moment comes, you wind up not with a gush of nectar but with a mouthful of meal.

What went wrong? Here are a few of the possibilities. In the first place, there is no such thing as a simple peach. There are hundreds of specific varieties, and some are better than others. Each comes ripe only for a brief period of the summer. Maybe what you got was a lesser variety. Or maybe it was picked too young. Like all fruits and vegetables, peaches accumulate sugar and flavor the longer they hang on the tree. There's also another wrinkle with peaches: they must ripen as well as mature, and this happens on a separate timetable. Even if your peach was picked at the right time, maybe it was sold before it was ripe.

So which peach should you have picked? Well, aside from buying from a farmer you know and trust — always the best choice — there are ways to buy a good peach even at the grocery store (and you'll probably be surprised at just how good a peach can be, once you know what to look for). That's what this book is for.

You don't know any of that? Don't feel bad. Most people — including chefs and even grocery store produce managers — don't either. Foodies now have no trouble at all explaining *brunoise* (and even pronouncing it correctly) or expounding on the differences between the Maillard reaction and caramelization. Yet many are completely in the dark about the very ingredients they're so expertly chopping and browning.

Not long ago I engaged in a brief online spat with a noted New York restaurant critic who was complaining in February about the terrible quality of the California strawberries at her market. "Lady," I felt like saying, "if you want to know who is to blame, look in the mirror." If you continue to buy out-of-season, out-of-state strawberries, growers will continue to sell them. Knowing how to choose, store and prepare fruits and vegetables is not a matter of magic or even of science. It's nothing more than learning to recognize a few clues and then paying attention.

We are living in an unequaled time of unparalleled plenty, even though it sometimes seems that the only thing American agriculture has not been able to deliver is that which is most basic — flavor. But things are changing. Finally, we're beginning to look seriously at where food comes from, how it's grown, how it's marketed and how it tastes. The successes of organic produce and farmers' markets — no matter how misplaced our faith in them sometimes might be — speak volumes about the desire for food that has flavor and that has been grown in a way we regard as wholesome.

Less commented on but perhaps more important, their successes speak to the willingness of some food shoppers to pay a premium for fruits and vegetables that meet those criteria. Even five years ago, after suggesting the possibility of farmers' markets to mainstream growers who were complaining about low prices, I would get the equivalent of a kindly pat on the head. Those markets were very nice for hippies, the growers would say, but they weren't for real farmers. Today I'm seeing some of those same guys selling their produce at the markets, and they seem happy as clams handling all that cash. (To be sure, all the growers at farmers' markets are not created equal. Even there you must be choosy about what you buy. It's just that the potential taste rewards are so much greater.)

· · ·

A WORD ABOUT ORGANIZATION

My selection of the fruits and vegetables in this book was dictated partly by the importance of the ingredient and partly by whim. I didn't include some vegetables and fruits — celery, for example — because I simply don't find them that interesting. I left out apricots because there aren't any good fresh ones left. Unless you happen to be at a very select farmers' market during the few weeks of the year when the local Royal Blenheims come in (if they come in), you're better off using dried apricots. (Most modern commercial apricot varieties were selected for drying anyway.) By contrast, I did include figs and persimmons, even though they can be hard to find, just because they are among my favorite fall fruits and I can't imagine doing a produce book without them.

Notice that seasons don't break cleanly on the calendar. Their occurrence varies according to location and weather. Whereas cherries are considered a spring crop in many places, they don't ripen until mid-July in others. And even in the same location, there may be as much as two or three weeks' difference from one year to the next.

When you start with good ingredients, you finish with great dishes. Furthermore, the better the raw materials, the simpler the recipes can be. Trying to coax flavor out of mediocre ingredients takes a lot of work (and, frequently, a lot of sauce). But start with something that already has flavor, and your work in the kitchen is mostly done. Braise it in a little butter, give it a squirt of lemon for backbone and there you are. You did hardly anything, but everyone thinks you're a genius.

Although trying to cook with the seasons can initially seem like a straitjacket, you'll soon find benefits you never suspected. It sounds paradoxical, but you'll find yourself cooking with more ingredients than you ever had before. One of the problems with having anything you want whenever you want it is that it is tempting to settle on only the few things that you know you like and never branch out. But why settle for a miserable peach in the dead of winter when you could have a wonderful mandarin, orange or grapefruit? By abstaining from those few favorite things except when they are at their very best, you'll discover a whole world of flavors that you may never have suspected existed. Even better, you'll find that when it is time once again to try those peaches, they will taste even better than you remembered.

An added benefit is that you'll be saving money. Transportation comes at a price, and so does the scarcity that is the nature of out-of-season fruits and vegetables. Wintertime peaches flown in from Chile not only have little flavor, but they also can cost the earth. When you buy them at the right time of year, however, when the local farmers have filled the markets with them, these fragrant treasures go for pennies. They'll even be cheap enough that you can afford to buy the very best. And that's the time you want to pick a peach.

The Plant Designers

FACTORIES IN THE FIELD

W hen you think of a strawberry, you envision a piece of fruit and wonder whether it will be sweet and juicy. When a farmer thinks of a strawberry, he sees a plant and has a completely different set of concerns. He wonders whether it will be resistant to all the many pests and diseases that afflict the species. He wonders whether it will have the vigor to grow strong in what might be challenging climatic or soil situations. Will it be a heavy bearer?

He wonders about the fruit, too. Will it appear at the times it makes him the most money — early and late in the season? Will it appear consistently, rather than in fits and starts, so he can keep a steady crew employed? Will it look good? (Supermarket shoppers buy with their eyes, not their palates.) Will it be firm enough to survive the days of shipping it takes to get to market? How will it taste? It has to taste good enough, certainly; that goes without saying. But if the rest of those criteria aren't met, all the flavor in the world won't mean a thing.

Plants that can do all those things do not just happen. Almost every fruit or vegetable that we eat is the result of centuries — even millennia — of careful breeding and selection. At its most primitive level, this happens almost accidentally. Because genetics is constantly evolving, seeds from even the same plant can grow into new plants that are slightly or even markedly different from each other — just like kids from the same set of parents. So farmers have always selected the seeds of their best plants every season and then stored them to start the next crop. In this way, the traits of the plant are slightly altered each generation. Eventually, this can result in fruits and vegetables that are almost wholly different from their ancestors. (The original corn plant produced ears only slightly bigger than those of wild grass.)

When practiced by family farmers working small plots of land, this kind of selection also can result in plants that are ideally suited for very specific growing areas. Botanists call these varieties "land races." For example, in the Veneto region of Italy, there are dozens — if not hundreds — of subtly different types of radicchio di Treviso, many adapted to a specific family's farm (or, since there can be differences in soil and climate within a single holding, even specific parts of a specific family's farm).

Sometimes these changes are neither slight nor gradual. New plants, radically different from their parents, can happen spontaneously. These genetic mutations, called "sports" by the trade, have played a much more important role in the development of modern agriculture than you might think. The Hass avocado was discovered this way in the 1920s, hanging from a tree of a different variety and then propagated. The original still stands in Whittier, California. The same happened with the Red Delicious apple. The story goes that an Iowa farmer gave the tree a try only after he failed in repeated efforts to kill it. And so, too, the Washington navel orange — the longtime industry standard. The original was a sport found in Bahia, Brazil, in the 1860s. In fact, the larger family of oranges may well have come about in a somewhat similar fashion, as there is no such thing as a true wild orange. (Botanists believe that they probably stem from crosses between giant pummelos and tiny mandarins.)

There was nothing accidental about the development of the modern strawberry. It required not only centuries of selective plant breeding but ocean navigation as well. The strawberry you buy in the supermarket today draws its heritage from very different plants found on entirely different continents. It developed spontaneously in France by a cross between plants from Chile and

North America. The last great leap in strawberry quality came in the late 1970s as the result of a happy accident. While walking in the Wasatch Range of Utah, legendary plant breeder Royce Bringhurst came across a wild strawberry fruiting at a time and place it shouldn't have been. He took it back to his lab and began using it in some of his crosses. The result was a strawberry that could grow in warmer weather and during longer days than was previously possible. These so-called day-neutral berries revolutionized the industry, allowing the eleven-month-a-year harvest we enjoy today. Although the genetics now seem to be more complicated than Bringhurst had originally thought, about 40 percent of the commercial strawberries in the country spring from these varieties.

The journey hardly stopped there, of course. The quest for a better berry continues today — and in much the same way it always has. On twenty acres of University of California–owned farmland tucked against the Irvine Hills in Orange County, strawberry breeder Kirk Larson is tending more than 25,000 plants. Somewhere among them, he hopes, the next great variety is waiting to be discovered. Larson doesn't know when it will turn up — it may be this year, or it may be five or six years from now. He does know that it will have to be something pretty special, as he and his predecessors have set the strawberry bar high.

In fact, it's not going overboard to say that in the past twenty years or so, plant breeders have revolutionized the strawberry. What once was a tender icon of spring has become a year-round supermarket staple. This has certainly been good for business: in 2003 strawberries became a $1 billion crop for California. The state now produces more than 85 percent of all the strawberries grown in this country and more than 20 percent of all the strawberries grown in the world.

But every revolution has its casualties, and in this one it was the sweet, melting, incredibly fragrant berry of generations past. You might have thought that berry was remarkable, but for the people whose livelihoods depend on supplying strawberries to the entire country, other characteristics are every bit as important. The history of strawberry breeding is littered with delicious berries that are no longer grown. The most recent example is the Chandler, which ruled the strawberry industry for almost a decade in the 1980s and 1990s until it was done in by its fragility. "The fruit looked great and tasted good, but if you shipped it to New York, when the buyer opened the box, the fruit was all leaking out the bottom — that's an automatic rejection," Larson says. "You've paid all that money to grow the fruit, pack it and ship it, and it's a total loss. That only has to happen a couple of times before you get real tired of it."

That may sound hard-nosed, but it is the philosophy that since World War II has transformed what was once a highly seasonal niche product into one of the marvels of modern industrial agriculture. Since 1950 California strawberry growers have increased their production more than twentyfold, from 81 million to 1.67 billion pounds. At the root of this transformation have been the University of California's strawberry breeders: Larson, working out of the South Coast Research and Extension Center in Irvine, and his partner Doug Shaw, who operates out of Davis in the north. UC varieties account for almost two thirds of all the berries grown in the state.

In addition to the University of California, there are two other private strawberry breeding programs in the state, managed by shippers Driscoll's and Well-Pict. These proprietary varieties make up about one third of the California harvest. Overseas, the UC breeders' impact is even greater. Roughly 90 percent of the strawberries grown in Spain, the second-largest producer, come from the University of California, and those varieties are similarly influential in strawberry fields from Mexico to Morocco.

That's the kind of success other plant breeders would love to have. To achieve it, they are taking plants to places they might not ordinarily go in nature. One of the biggest pushes in fruit breeding is trying to find seedless varieties. Obviously, left to their own devices, seedless fruits would not last long. But they are incredibly convenient to eat (witness the almost total domination of seedless grape varieties today and what that has done for grape sales). Seedless fruits do occur in nature, but only under very specific conditions and, of course, only for one generation. In some plants, the development of fruit occurs separately from pollination. Some fruits don't produce seeds because of a genetic imbalance. Bananas, for example, have an uneven number of chromosomes, making the production of seeds rare. (It does happen, though, and you'll sometimes see a little train of tiny black specks going up the center of the fruit.) Other good examples of this are navel oranges and Satsuma mandarins, which also almost never have seeds. Other fruits will not produce seeds when they are pollinated only by plants of the same variety. Plant Clementine mandarins and Minneola tangelos in orchards that are isolated from other citrus, and they'll be nearly seedless; plant them close enough to other citrus, and they can be quite seedy. Oddly enough, seedless grapes are a different matter, because they are not truly seedless. Instead, these grape varieties produce seeds that never mature beyond the size of a speck. Technically, they have seeds, but you'd never notice.

So how do seedless plants survive? Mainly with the help of farmers, who propagate them by cutting branches and grafting them onto other plants, or by encouraging the cuttings to develop roots of their own. (In nature this can occur when a plant produces side shoots, which will continue to live after the infertile shoot dies off.)

Seedless mandarins — more commonly known as tangerines — are the primary focus of citrus breeders today. Mandarins are terrifically convenient to eat because they peel so easily. On the other hand, they usually are loaded with seeds (Satsumas and Clementines being the main exceptions). In addition, peak mandarin season is right around Christmas, with a few varieties extending into February. The breeder who can successfully combine easy peeling, seedlessness and an extended market season will have found the holy grail of citrus: a fruit that can be eaten as conveniently as a candy bar.

That breeder might be Mikeal Roose and his crew at the University of California at Riverside. In 2002 Roose introduced three new seedless mandarin varieties, with the California-centric names Shasta Gold, Tahoe Gold and Yosemite Gold. They are all seedless and easy to peel, but beyond that they have marked differences. Shasta has rich, intense mandarin flavor, matures very late and will hang on the tree well into May. Tahoe is earlier but has the same rich flavor. Yosemite falls somewhere in the middle as far as harvest and has an almost candied, Kool-Aid flavor. So far, growers have been slow to pick up on them. The mandarin industry in California is just beginning to catch fire, and it's far more expensive to graft over an orchard of perennial citrus trees than it is to replant strawberries, at best a biennial.

And, to tell the truth, there haven't been a great many new citrus additions in the past one hundred years or so, and those few haven't fared awfully well. Probably the last great citrus fruit rollout was the Oroblanco, a cross between a pummelo and a white grapefruit. It is a spectacular fruit when well grown, sweet but with the pummelo's characteristic bitterness to balance. Introduced in 1980 to much fanfare, it was picked up by growers pretty quickly.

But the Oroblanco has an unfortunate quirk: its peel is often quite thick and sometimes keeps a greenish cast well past the time the fruit is ready to eat. This discouraged shoppers, and gradually the fruit fell from favor. There's a twist to the story, though: Israeli farmers picked it up, renamed it the Sweetie and marketed it in Japan as a green fruit. The Japanese loved it, too, and now, twenty-five years after its introduction, Oroblanco plantings are again on the upswing in the United States.

When you walk through Riverside's research orchards — a pair of tracts totaling twenty-two acres in the foothills next to the university — both the promise and the perils of citrus marketing are evident. Established in 1912, they now contain more than nine hundred types of citrus. Not only are new trees tried out, but the orchards also contain a substantial collection of historical varieties — many of which are no longer grown — and types of citrus that are popular in other countries but have not been introduced here.

Among the latter is an Australian finger lime — a greenish black, long, thin fruit that is filled with lime-colored and lime-scented beads that pop in your mouth like so much citrusy caviar. There's also something called a "dawn fruit" — a cross between a blood orange and another pummelo-grapefruit combination similar to the Oroblanco. It is absolutely gorgeous, with deep purple flesh and a compelling raspberry-tinged flavor. As good as it is, it will quite likely end up going nowhere, as the costs of introducing a totally new product into the market are staggering.

That's the main reason so few fruits and vegetables are labeled by specific variety in the market. It's much easier to sell a "peach" than it is to introduce shoppers to the half-dozen new types that appear every year. This is one more area in which farmers' markets are changing things, though. More and more, varietally labeled produce — everything from Chioggia beets to Pixie mandarins — is gaining cachet and selling for premium prices.

The promise of varietal marketing is the last thing on Larson's mind as he prowls his trial fields examining his brood. His fruit will be sold simply as a strawberry, which is a blessing in some ways but totally ignores all the work that went into developing it. To understand how remarkable these plants are, you have to understand the strawberry itself. If ever there was a fruit that seemed designed to frustrate industrial agriculture, this is it.

From the tips of its berries to the ends of its roots, the strawberry is the essence of vulnerability. Unlike most fruits, strawberries wear their seeds on the outside of their flesh rather than the inside, and there is neither skin nor rind to protect the sweet flesh from the predations of climate and pest. This fragility is only amplified after the fruit has been picked, as the teeniest nick or cut can become an open doorway for the fungi that lead to rapid decay. If anything, the roots are even more exposed than the fruit. They are made of extremely simple tissues and lack any corky bark or covering that would protect them from soilborne pests. As a result, an established field of strawberries is little more than a well-advertised, all-you-can-eat buffet for almost any bug around.

Because of all this, strawberries are extremely expensive to produce. A recent study by the California Strawberry Commission, a growers' organization, predicted costs of more than $30,000 an acre to grow the fruit, harvest it and ship it. Farmers are left scrambling for any advantage to make their investment pay off.

That's where Larson and Shaw come in. Their job is to find the one plant in a million — almost literally — that can give farmers an edge. It might be a berry that ships better, or one that produces higher-quality fruit. (Even with the newest best varieties, farmers discard up to 20 percent of their fruit as too small, poorly colored or misshapen.) Or it might be a plant that produces exceptionally well either early or late in the season, when the prices are highest. The same flat of strawberries that sells for $6 in July may well go for $30 in January or February. But whatever special traits it has, it will also have to have good flavor, because without flavor the market won't be interested in it.

The search for this perfect plant begins not with any kind of high-tech genetic manipulation but with old-fashioned birds-and-bees pollination. Larson selects a pollen-bearing male flower from a plant with desirable traits and rubs it around the center of a female flower from another selected plant. When that fruit ripens, the seeds will be collected and planted. (Technically, these "achenes" are actually tiny dried fruits themselves, each one containing a seed.) Each plant will be a fraternal twin of the others from the same berry, sharing many traits, though also being slightly different. The plants that seem most promising will be selected and planted outdoors the next year.

In the field, the new plants are numbered and grouped according to their parentage. If you look closely enough, you can see the resemblances and also the differences. This plant may have large, perfectly formed fruit, while its sister right next to it bears strawberries that are of a similar size but more frequently misshapen.

Most of what Roose does in his citrus breeding program is traditional genetics as well, although he is experimenting with exposing cut branches of trees to low levels of radiation to induce mutations that might result in favorable new characteristics rather than simply waiting for them to happen. This is about as adventurous as most fruit and vegetable breeding gets. Although there has been widespread public concern about genetically modified organisms (GMOs), they are largely restricted to commodity crops such as field corn, cotton, canola and soybeans.

The major exception was the vaunted Flavr Savr tomato, one of the first GMO products. Introduced with great fanfare in 1994, it was supposed to be able to sit on shelves forever without going soft. Unfortunately, once shoppers found that it tasted like cardboard, sitting on shelves was about all it did. Combined with the fact that the plant proved uneconomical to grow in the areas where it was planted, the experiment was a bust, and the Flavr Savr was quietly dropped in 1998.

In most cases, this kind of genetic modification is done to make plants resistant to pesticides and herbicides, so they can be sprayed with a certain company's chemicals without suffering any damage. This practice may be a boon to growers, but only to growers who buy both the seeds and the chemicals from the same company. No ill effects to the consumer have been demonstrated, but the lack of evidence is hardly reassuring, since the Food and Drug Administration has decided that testing is not required in most cases. Furthermore, the industry is fighting tooth and nail against labeling GMO products, which would give consumers a choice as to whether they want to buy them. Ironically, probably the most GMO-intensive food you can find today is tofu, as about 70 percent of American-grown soybeans are genetically modified.

The work of traditional plant breeding is unimaginably

tedious. Larson walks his strawberry fields nearly every day for the twenty-eight weeks that make up the typical Southern California season. He uses a handheld Allegro CX computer loaded with a spreadsheet to score each plant on a variety of criteria. He counts the fruit and grades it for size, shape, color and, to a certain extent, flavor. He grades the plant's "architecture" — how the leaves and fruit are presented — and several other factors. "I'm looking for the drop-dead gorgeous strawberry poster child," he says.

At least in the early going, he doesn't pay as much attention to taste. He tastes the berries from time to time, but flavor is too subjective and complex. "There are so many aspects to strawberry flavor," he says. "It's not just sweetness. There has to be the right blend of sweetness and acidity, and then there are all the aromatics that make up flavor as well." Of course, there are more mundane reasons for not tasting every berry. "I do love strawberries, but I'd have to be eating thousands a day to make it meaningful," he says. "I try to take a bite whenever I can from plants I like, at least until I start feeling bloated." Flavor will become much more important at the end of the plant's selection trials.

At the end of the year, Larson runs the spreadsheet and analyzes each plant for productivity and value. Of the 13,000 or so plants in the initial selection, he picks the best to go on to the next year, based on the criteria of shape, size, color, firmness, architecture, disease resistance, rain tolerance and flavor. This time he cuts runners from the strawberry and plants them. (Seeds vary genetically, whereas propagating from plant division ensures identical clones.) The process will repeat with further winnowing each year until one plant has proved its worth to become a variety. This season, in addition to the new trials, 487 strains from last year's first selection made the cut. There are also 160 from the year before that, and a few from even earlier. One plant survived more than five years, but Larson says this year will be its last. "I had pretty high hopes for that one to become a variety, but it just doesn't look like it's going to make it," he says. It had great yields, early production and excellent shelf life, but the flavor just wasn't consistent.

Even after a variety has been selected, it still has to be tried out in test plots maintained by farmers at various locations throughout the state. All told, it takes at least six years to find a plant that is worthy of becoming a named variety. It frequently takes even longer. The Chandler berry, which was introduced in 1983, took seven years for Larson and Shaw's predecessors, Victor Voth and Royce Bringhurst, to develop. And it took Voth a decade to select the Chandler's successor, the Camarosa.

Still, that's a dead sprint when compared to developing new citrus. A fruit tree takes three to five years to reach productive maturity (and eight or nine to hit its peak), not a single season like a strawberry plant. Once a tree with potential has been selected, it goes to an experimental station in the San Joaquin Valley to be tried out there and at other locations as well, since the citrus industry is dispersed through several areas with different climates and soil types. That takes another three to five years. Then it's released to commercial orchards, where it is tried out again. By the time you add it all up, developing a new citrus takes an average of about thirty years.

It is rare that any breeder sees a citrus tree through from beginning to end. Rather, breeders operate like stonecutters on medieval cathedrals, inheriting works in progress from the previous generation and passing on their own projects to the next. The Gold series of mandarins that Roose introduced in 2002 was begun in 1973 by his predecessors, James Cameron and Robert Soost. They in turn introduced varieties such as the tiny, seedless Pixie mandarin, which had been originally developed in the 1920s by Howard Frost, who along with Walter Swingle practically invented modern citrus breeding. The profession can make for some testy relationships. Larson says he never really understood why Voth was so prickly until a friend explained the difficulty of passing along plants you have developed and nurtured: "Hey, it's like he was married for fifty years, and all of a sudden he wakes up and you're sleeping with his wife."

Once a variety is chosen and endorsed by the growers, it can spread quickly. A single citrus tree can provide thousands of buds for grafting onto other trees. Within four years of their introduction, there were more than 55,000 Yosemite Gold, Shasta Gold and Tahoe Gold trees planted in California orchards.

As you might expect, progress is even faster with a plant like strawberries, which are usually replanted every year. Within just a couple of years of its introduction in 1993, the Camarosa accounted for all but a tiny fraction of the strawberry acreage in Southern California and a huge chunk in the northern part of the state as well. Now the Camarosa is well on the wane, down to less than 20 percent of the state's harvest. Its successor, the Ventana, was first planted commercially in 2002 and within three years accounted for almost 15 percent of the harvest.

Each succeeding variety offers improvements on the previous one but also has flaws of its own. "No variety is perfect; there are always trade-offs," Larson says. "There are one hundred criteria we have to assess, and there is no perfect variety that has all of them — and there probably never will be.

"In the twenty years since the Chandler was introduced, we've pretty much doubled the yield and probably tripled the shelf life of strawberries. We've added at least another month to the season, if not six weeks. If this were a new biotech start-up, everyone would be all over us. But because it's strawberries and we're just sticking pollen from one plant onto another, something that's been done for thousands of years, nobody thinks much about it. Growers are impatient. Everybody always wants something new. We just released the Ventana a couple of years ago, and already people are asking me when I'll have a replacement for it. I'm telling you, it's not easy to do, at least partly because we've set the bar so high. If you wanted to replace the Chandler right now, I've got hundreds of plants that would be better. Replacing the Camarosa or the Ventana is not so easy."

The Chandler berry had great flavor when everything went well, but it was too soft to ship. The Camarosa, which is exceptionally firm and ships very well, has much larger fruit, is a heavy bearer and is much easier to harvest, but the flavor isn't quite as good as the Chandler's. The Ventana bears a lot of fruit early; the fruit tends to have better shape, texture and color than the Camarosa; and it is at least equal to the Camarosa in flavor.

Along with the successes have been varieties that didn't work out quite as well as the breeders had hoped. The Gaviota, for example, which was introduced in 1997 and is doing well in the farmers' market niche, has excellent flavor but does not bear heavily, and the fruit doesn't last. The Diamante, also introduced in 1997, has good flavor but is firm and lacks internal color, which limits its use in processing. Growers will probably begin to replace it when the new Albion variety is ready for the big time. "That's going to knock people's socks off with flavor," Larson says.

That remains to be seen. But you can be pretty sure that however it tastes, this brave new berry will be big, pretty, easy to pick and prolific; will bear over a long period; and will be firm enough to ship from here to the moon without suffering a bruise.

Spring

Artichokes

Alexander Pope wrote that it was a brave man who first ate an oyster. What possible words can describe the heroism of he who first ate an artichoke? Not only did he have to consume it, but he probably had to invent it as well. At first glance — and maybe even after patient consideration — little about the artichoke indicates either edibility or conscious creation. The thing looks more like a primitive instrument of war than a domesticated product of agriculture. With its overlapping rows of hard prickly petals, it seems only one step removed from a stick with a nail stuck in it. Yet somehow, sometime, someone almost certainly did create the artichoke. Exactly how, when and who are unclear. Obviously, it happened well before anyone thought to copyright a plant, or even to write a scientific paper claiming academic bragging rights. But there is little doubt that the artichoke was invented.

The vegetable that we call an artichoke is actually the unopened flower bud of a plant that is an improved cardoon. (My colleague Charles Perry says the word "artichoke" is derived from the Arabic *al'qarshuf,* which translates as "little cardoon.") If you visit ethnic produce markets — particularly Italian ones — you may have seen a cardoon. It looks like a prehistoric stalk of celery. It is outsize and a pale dinosaur gray-green with a thick, stringy skin. Peel it, chop it and cook it, and you'll taste artichoke.

Why did our unnamed farmer decide that the bud of the cardoon was more desirable than the stalk? Is that even what he was going for? Did he really think he had accomplished his goal, or did

he simply give up? There is something haphazard, even accidental, about the artichoke. One thing's for certain: no modern plant
breeder would dare to come up with something like it. More's the
pity. The artichoke is one of spring's great vegetables, with a buttery texture and an appealing flavor — an almost brassy sweetness
that combines well with a multitude of other ingredients.

But there's no getting around it, the artichoke is a peculiar vegetable. First, of course, there is its form — like a thistle-covered
mace. The edible part of the artichoke is an unopened flower bud,
or, more accurately, a collection of flower buds. If it is left to open,
the artichoke will turn almost inside out, blossoming into something that looks like a flat pincushion stuck with hundreds of tiny
lavender-blue flowers. It is attractive in its own gargantuan way,
and fully opened artichoke flowers are sometimes used by avant-
garde florists to make visual statements in arrangements. The
sharp, tough "petals" or "leaves" of the artichoke are what botanists
call bracts, which are actually somewhere between the two. Bracts
are tough, leaflike objects that protect the flower.

But the artichoke's contrariness is more than skin-deep. In fact,
peel an artichoke and set it aside for a minute, and you'll soon discover another of its eccentricities. Exposed to air, artichokes turn
brown or even black. This is not altogether unusual in itself — potatoes do the same thing, and so do peaches and shrimp, among
many diverse foods.

The process is what chemists call enzymatic browning. The plant
contains a substance that when exposed to oxygen changes the
color of the flesh. This is not always bad. All tea would be green if it
were not for enzymatic browning. In the case of artichokes, though,
it's hard to see the benefit, at least for the cook. But whereas it is
almost impossible to prevent enzymatic browning, we can delay it
fairly easily, either by preventing exposure to oxygen or by treat-
ing the flesh with an acidic compound. Neither of these takes any
special equipment, just a bowl filled with acidulated water — plain
old tap water to which you've added an acid of some sort (white vinegar and lemon juice work equally well). When you're done, keep
the artichokes in the water until you're ready to cook them. Old-
time chefs used to call for cooking artichokes *en blanc* — in a com-

bination of water, acid and flour. This only slightly improved the color and pretty much wrecked the flavor for anything other than serving them as glorified chips and dip. You're better off settling for only minimal browning.

Another odd thing about the artichoke is its tendency to make everything taste sweeter — not in a good way, but that weird metallic kind of sweet you get from diet soft drinks. This is mostly caused by a naturally occurring chemical called cynarin (artichokes belong to the genus *Cynara*), which is unique to artichokes. This sweet reaction can be so powerful that it is almost off-putting. Sometimes the flavor is so strong that even a sip of water tastes as if it has been artificially sweetened. It is no surprise that this sweetening makes artichokes extremely unfriendly to wine. It can be reduced by extended cooking, which results in a gentler, more complex flavor. Remember that when you're thinking about a dish: Cook artichokes briefly, and they will have a big, brassy edge that can stand up to the most aggressive seasonings — anchovies, garlic, black olives . . . bring 'em all on. Cook the vegetable more gently, and you'll be surprised at its delicacy.

Unlike most vegetables, which can be harvested only during a single season, artichokes actually bear twice. There is a large harvest in the spring — March to May accounts for about 70 percent of the total crop — and then a smaller one in late fall. Some connoisseurs claim to be able to detect a difference between spring and fall harvests, but if there is one, it is incredibly slight.

And, as if these weren't enough oddities for one plant, the artichoke comes in many different sizes. In season the so-called baby artichokes can be one of the best buys in the produce department. These are actually fully mature chokes that are harvested from exactly the same plants as the big boys at exactly the same time. An artichoke plant sends up many flower stalks, some as tall as six feet. One or two of them will yield the large, steamer-size buds (weighing a pound or more apiece). Maybe half a dozen of them will be medium-size chokes (two or three to a pound). And then there will be a scad of smaller ones (roughly a dozen to a pound). Because most shoppers are interested in artichokes only for steaming, these smaller ones are tough to sell. Most of them go to canning, but many of them wind up in the produce aisle, where they're sold cheap to savvy cooks who know their true value.

WHERE THEY'RE GROWN: Almost all of the artichokes in the United States are grown in California, most of them within fifteen miles of a small town called Castroville. There have been recurring efforts to expand the plantings to other areas in order to expand the season, but they have met with only mixed success.

HOW TO CHOOSE: Artichokes are one of the tougher vegetables; they'll last quite a while with only minimal care. Still, choose the ones that seem heaviest for their size and that don't have any visible damage. You don't have to be too picky about this: the cut stems will, of course, be blackened already. And if there are a few dark spots, they won't affect the flavor. The industry has come up with the marketing term "frost-kissed" for this kind of damage and claims that it makes the hearts sweeter. Perhaps, but it certainly doesn't hurt them any. You can tell really fresh artichokes because their leaves will squeak when you rub them together.

HOW TO STORE: Keep artichokes in the refrigerator, tightly sealed. Don't clean them until shortly before you're ready to cook them.

HOW TO PREPARE: The big "hubcap" artichokes that sell at such a premium price should be steamed, boiled or microwaved. You can eat them leaves and all. To clean them, cut the stem off flush with the bottom so the artichoke will sit upright on a flat surface. Tear off the tough outer ring of leaves, bending them back from the choke until they snap. Then pull down — this will tear away the worst of the tough, stringy outside. Use sharp kitchen scissors to cut away the top third of each leaf — the spiny part. Rub the cut surfaces with lemon. Steam the artichoke until it is tender. Exactly how long you need to cook it will depend on the size and the age of the artichoke — figure anywhere from 10 to 30 minutes. The artichoke is done when you can easily pull a leaf free. I like to remove the bristly heart before serving. To do this, spread the top center leaves as wide as possible without breaking them, then use a serrated grapefruit spoon to dig out the choke.

ONE SIMPLE DISH: The best way to prepare artichokes is by braising. This method is remarkably easy and flexible. Here's the general outline: Put 2 pounds of artichokes that you've trimmed well in a large skillet with a couple of tablespoons of olive oil, ½ cup of water and a minced garlic clove. If you like, add some red pepper flakes. Cover the skillet and let simmer over medium heat until the artichokes are tender, about 5 minutes. Raise the heat to high, remove the lid and cook the artichokes until most of the moisture has evaporated and what remains has emulsified with the oil. Toss the artichokes in this glaze and serve immediately.

HOW TO PARE AN ARTICHOKE

You prepare artichokes for cooking differently from when you are planning to stem them and eat the leaves. The goal is to wind up with just the edible parts — the softer inner leaves, heart and stem. You can do this by tearing away the outer leaves by hand, but the following method is much faster. Since you'll usually be paring at least half a dozen for most recipes, this is a technique worth learning.

TO BEGIN, have a large bowl at your side filled with water and the juice of half a lemon. This acidulated water is where you will put the cleaned artichokes; the lemon juice will keep them from discoloring.

HOLD THE ARTICHOKE in your left hand with the stem facing toward you and the tip facing away. Slowly turn the artichoke against the sharp edge of the knife while making an abbreviated sawing motion. (It's easier to control the knife if you use the base rather than the tip.) You will begin to cut through the tough outer leaves; when you can discern the natural cone shape of the artichoke, adjust the knife to follow it. Keep trimming like this until you've cut away enough of the tough leaves that you can see only light green at the base of the leaves. Cut away the top inch or so of the tip of the artichoke, then dip the artichoke into the lemon water to keep the cut surfaces from discoloring.

WITH A PARING KNIFE, trim away the very tip of the stem, then peel the stem and base of the artichoke, going from the tip to where the base meets the leaves. You'll have to do this in five or six passes to make it all the way around the artichoke. When you're done, there should be no dark green tough spots left, only pale green and ivory. If you're using baby artichokes, leave the choke whole. Just put it in the lemon water and repeat the instructions for the remaining artichokes.

AFTER YOU'VE PARED THE ARTICHOKE, if you're using medium-size ones, you'll probably want to quarter it. The easiest way to do this is to set the artichoke on the cutting board so it is upside down, resting on the cut surface of the tip. Cut it in half vertically, then in half again. Check the choke. With most, there will probably be nothing but a little fuzz; you can leave that and just put the quarters in the lemon water. Larger artichokes will have what looks like very fine hair. Cut just below that to the very base of the leaves, and the base will pop off, leaving a clean heart below. Put the cleaned quarters in the water and go on to the next artichoke.

Artichokes Stuffed with Ham and Pine Nuts

Baked this way, artichokes turn almost silky, while the stuffing browns to a nice crust. Try experimenting with this stuffing; chopped green olives are good, too.

6 SERVINGS AS A FIRST COURSE

2 tablespoons pine nuts

½ lemon

6 medium artichokes

1 garlic clove

¼ baguette, crust trimmed and cubed (about 2 ounces)

2 ounces ham, cubed

3 tablespoons chopped fresh parsley

1 teaspoon grated lemon zest

½ teaspoon salt

1 cup dry white wine

Toast the pine nuts in a small skillet over medium heat, stirring, until they are lightly browned and fragrant, about 10 minutes. Set aside.

Heat the oven to 350 degrees. Fill a bowl with water and squeeze the lemon juice into it. Keep the squeezed-out lemon half.

Pull off the lower leaves and tough outer leaves from an artichoke; this will be about the first two rings. Using kitchen shears, trim the top half of the next several rings of leaves until you get to the tight central cone, where the leaves are pale green at least two thirds of the way up. Use a knife to cut off the dark green top third. Trim the stem of the artichoke with a paring knife, making a flat base. Rub all the cut surfaces with the lemon half.

Place the artichoke upside down on a work surface and press firmly. Turn the artichoke right side up and use your fingers to spread the leaves as much as possible without breaking them. Use a grapefruit spoon or other small spoon to remove the innermost purple-tipped leaves and then scrape the fuzzy choke from the base.

Place the cleaned artichoke in the bowl of lemon water and repeat with the remaining artichokes.

When all of the artichokes have been cleaned, mince the garlic by dropping it down the feed tube of a food processor while it's running. Stop the machine and add the bread cubes. Pulse 2 or 3 times to break these down. Add the ham and parsley and pulse until the bread and ham are in large crumbs, 4 or 5 times. Remove the blade and stir in the pine nuts, lemon zest and salt.

Drain the artichokes and arrange them in a baking dish just large enough to hold them in a single layer. Fill the central cavity of each artichoke with some of the stuffing mixture, mounding it over the top and working a bit of it between the leaves.

Pour the white wine into the bottom of the baking dish and add just enough water to come to a depth of about ¾ inch. Cover the dish with aluminum foil and bake until the artichokes are tender enough that you can easily pull out one of the interior leaves (a knife will pierce the base easily as well), about 40 minutes.

Remove the baking dish from the oven and carefully pour the leftover liquid into a small saucepan. Bring it to a boil over high heat and reduce to a thin syrup. Pour the syrup over the cooked artichokes and set aside to cool to room temperature before serving.

Cream of Artichoke Soup with Parmesan Chips

A velvety cream soup using only ¼ cup of cream? The trick is adding a potato, which not only contributes a subtle earthy flavor but also adds enough starch to thicken the soup. Be careful with these Parmesan chips (a version of the northern Italian classic *frico*): if left unattended, they tend to disappear all on their own.

8 SERVINGS

- 4 medium artichokes (about 2 pounds)
- 1 small onion, diced
- 3 tablespoons olive oil
- 1 carrot, diced
- 4 garlic cloves, minced
- 1 medium boiling potato (about ½ pound), peeled and diced
- Salt
- 4 cups chicken broth
- ¼ cup heavy cream
- 2 large egg yolks
- Fresh lemon juice (optional)
- Parmesan Chips (recipe follows)

Pare the artichokes as described on page 40, then cut into quarters and keep in a bowl of acidulated water until ready to use.

Place the onion and olive oil in a large saucepan over medium heat. Cook until the onion begins to soften, about 5 minutes. Add the carrot, reduce the heat to medium-low, cover and continue cooking until the carrot has softened, 5 to 10 minutes, being careful not to let the vegetables brown.

Add the garlic and potato and cook for another 5 minutes. While the vegetables are cooking, chop the artichoke quarters into ½-inch pieces and add them to the saucepan. Add 1 teaspoon salt and cook, covered, for 10 minutes to begin marrying the flavors.

Add the broth and bring the mixture to a simmer over medium-high heat. Reduce the heat to medium-low and cook until the vegetables are soft enough that you can smash them between your fingers, 25 to 30 minutes.

Transfer the mixture to a large bowl and wipe the saucepan clean. Puree the soup in a blender on high speed until light and smooth, about 30 seconds. Do this in 3 batches to keep from overflowing the blender. As each batch is pureed, pour it through a fine-mesh strainer into the saucepan. Stir the soup that's in the strainer to help it pass through.

When all of the soup has been pureed and strained into the saucepan, return the saucepan to medium-low heat. In a bowl, whisk together the heavy cream and egg yolks until smooth. When the puree has neared the simmer, spoon ½ cup of it into the cream-egg mixture and whisk until smooth. This will "temper" the egg yolks, cooking them slightly so they won't curdle when you add them to the saucepan.

Slowly whisk the cream-egg mixture into the saucepan and continue to cook, whisking frequently, until the soup has thickened slightly and the texture has become silken, about 5 minutes. Do not let the soup boil, or the egg will curdle. Season to taste with salt and lemon juice, if desired.

Ladle about ½ cup of hot soup into each of eight small soup bowls and garnish with a crisp Parmesan chip in the center. Serve.

Parmesan Chips

MAKES 8 CHIPS

¾ cup freshly grated Parmigiano-Reggiano

Place the cheese in 1½–tablespoon mounds in a large, dry nonstick skillet. Leave plenty of space between the mounds; you'll be able to cook 4 or 5 mounds per batch. Gently spread each mound into a thin round with your fingers. Place the skillet over low heat. The cheese will melt, begin to bubble and finally, after 3 to 5 minutes, begin to brown around the edges. Sliding the tip of a fork under one edge, flip each chip and continue cooking until the other side is lightly browned, 3 to 5 minutes.

Artichokes Braised with Saffron, Black Olives and Almonds

Artichokes can stand up to the most assertive of flavors. This recipe pushes that theory to the limit, with orange zest, black olives and red pepper flakes (to say nothing of almonds and parsley). The addition of saffron makes the whole dish sing.

4 SIDE DISHES, 2 MAIN DISHES

1¾ pounds medium or 2¼ pounds baby artichokes, pared (see page 40) and placed in acidulated water (if using medium artichokes, quarter them lengthwise)

 1 tablespoon minced garlic

 ½ teaspoon finely grated orange zest

 ½ teaspoon salt

 Dash red pepper flakes

 Dash saffron

 ½ cup water

 ¼ cup olive oil

 1 tablespoon fresh orange juice

 2 tablespoons chopped, pitted oil-cured black olives

 1 tablespoon slivered almonds

 1 tablespoon minced fresh parsley

Cook the artichokes, garlic, orange zest, salt, red pepper flakes, saffron, water and olive oil in a large, covered skillet over medium heat until the artichokes are tender, about 15 minutes. Shake the pan from time to time to stir the contents. If necessary, add a little more water to keep the bottom covered.

When the artichokes are easily pierced with a knife, remove the lid and raise the heat to high. Cook, stirring, until only a thin coating of liquid remains on the bottom of the pan, about 3 minutes.

Add the orange juice, olives, almonds and parsley and cook, stirring, just until they are heated through. Serve warm.

Asparagus

Traditionally, it's the egg that represents spring's promise of rebirth. A much better candidate would be the asparagus. In the first place, it tastes better. Can you imagine anything more enticing than a big platter of fat spears, boiled or steamed just to the perfect point of mousse-iness and then drizzled with nothing more than a mild olive oil and fresh lemon juice? In a just world, that feast would be as required for the first day of spring as a roast turkey is for the heart of fall.

The cult of asparagus lovers — though still somewhat small when compared to, say, the mass religion that surrounds the summer tomato — tends to be no less passionate. It is, however, much more deeply divided, with many competing sects that are willing to go to the mat over various elements of doctrine. Asparagus lovers will argue about the merits of fat or thin spears. They'll debate the necessity of peeling. They'll quibble over methods for removing the spear's tough base. A tiny group of the truly orthodox holds out for the supremacy of the white asparagus. Fans will even argue about their beloved vegetable's effect on their urine.

Let's take a closer look at these issues. Believers in skinny spears claim that their asparagus is superior because it doesn't need to be peeled. It is delicate and crisp. Its very slenderness is evidence that it truly is the first spear of spring. Followers of the fatties respond equally dogmatically. Their spears aren't tough at all, they protest. Further, they are obviously the one true choice because what could be more fitting for the promise of spring than a rich, juicy texture?

In reality, they're both right. Skinny or fat, all asparagus is good in its own way. You just need to know how to use it.

First, a slender spear is not a sign of the first harvest. In fact, more often than not, the opposite is true. Whether asparagus is thick or thin depends on many things, but among them is what farmers call vigor — how healthy the plant is. On this issue the scientific evidence is clear: plants just beginning to produce make fatter spears. It's not as cut-and-dried as that, though, because the same plant will produce a whole range of sizes. Asparagus grows from a mass of roots, and each mass sends up scores of spears. Those that come up closer to the center, where the plant stores its nutrients, are fatter. Those farther out on the fringe are thinner.

Furthermore, although fatter asparagus does have a thicker, more fibrous peel that needs to be removed before cooking, it also has much more of the tender inner flesh as well. The peel is thinner on slender asparagus, so it doesn't need to be removed, but the juicy center portion is smaller, too. With fat asparagus, the peel is thicker and more fibrous toward the bottom. So start peeling from the tip, using gentle pressure, and then gradually increase the pressure toward the base. This will get rid of all the tough parts and leave only the juicy core. On a related plumbing matter, no matter how big of a hurry you are in, do not put those asparagus peels down the drain. Even if your garbage disposal will grind glass, it will not break up asparagus peels.

Whether spears are fat or thin, you'll almost always need to trim their bases (the only exceptions being if you buy asparagus that has already been cut — some stores do this for you). The lower portion of all asparagus is tougher and stringier than the tips — so tough and stringy, in fact, that it can't comfortably be eaten as it is. The conventional wisdom for removing the tough part is to hold the tip of the asparagus in one hand and the base in the other and bend the spear until it snaps. This wastes a lot of asparagus. Try it yourself: snap some asparagus and then cook the supposedly woody end. You'll find that much of what you would normally discard is edible — even delicious — down to the last inch and a half or so. You're much better off simply trimming that portion with a knife. Don't be too quick to discard the bases, especially if you're preparing a lot of asparagus. They may be stringy, but they do have good flavor. Cook them in broth, puree them in a blender and then run them through a strainer, and you'll have a very nice creamy asparagus soup. If you're making a risotto with asparagus (something I highly recommend), save both the bases and whatever peels you have. Add them to the simmering broth, and in 15 to 20 minutes it will be infused with asparagus flavor.

The choice of skinny or fat comes down to whether you're going to use the asparagus as a vegetable or as an ingredient. On the one hand, if you're thinking of a plain platter stacked high with asparagus that has been simply boiled or steamed and lightly dressed with, say, a brightly colored sauce mimosa, the only choice is thick spears. And if it's the first asparagus of the season, there should be so much that it makes up the whole dinner. Properly cooked — which is to say long enough that the asparagus is tender all the way through but quickly enough that the flavor and color haven't begun to fade — fat spears are incomparably rich and juicy. The texture is almost like a vegetable mousse, with just a slight resistance to the tooth. On the other hand, if you're planning a dish that will include asparagus along with other ingredients — a risotto or a frittata, for example — go ahead and pick the thinnest spears you can find. They'll combine better with the other elements in the dish, whereas the thick ones might dominate. Since it hasn't been peeled, skinny asparagus also tends to be a little crisper and have a brighter color that stands out better in a mix of ingredients.

As far as white asparagus is concerned, it's hardly worth arguing over, since so little of it is available in the United States these days, and what little you will find is usually fabulously expensive. White asparagus isn't a different variety, but rather spears that have been meticulously grown to prevent any exposure to sunlight. Sunlight creates chlorophyll, which has a green color. To grow white asparagus, farmers repeatedly pile more and more soft earth over the asparagus tips just as they begin to emerge, thus blocking the light. If you do find some white asparagus and decide to splurge, be very careful when peeling it. To get to the tender heart, you'll need to peel more deeply than you do for green asparagus, and because it is so expensive, you'll want to avoid as much waste as possible. In Germany, Switzerland and Austria, where white asparagus is celebrated, traditionally only the head chef is allowed to peel it, and only with a special vegetable peeler with an adjustable blade — designed just for that job.

One more controversy persists over asparagus, and those with delicate sensibilities must be cautioned. For centuries some people have noticed a change in the odor of their urine after eating asparagus, the result of the body's metabolizing sulfurous compounds in the vegetable. (Proust credited asparagus with "transforming my chamber pot into a vase of aromatic perfume.")

Until relatively recently, scientists were divided into two camps on this matter. One group held that some eaters are genetically predisposed to produce a chemical when they eat asparagus. The opposing group believed that some people have a genetically linked ability to smell the chemical. Today it is more commonly thought that the second theory is correct. To test this theory, scientists in the 1980s made a crucial change in their experimental regimen: rather than having people test their own urine for smell, they began randomizing samples and testers.

On a more savory note, one thing almost everyone can agree on is the importance of not overcooking asparagus. When overcooked, asparagus not only changes color from a gorgeous jade and emerald to a woeful olive drab. It also produces a chemical called methoxypyrazine, which has the distinct aroma and flavor of canned asparagus. There may be some debate, however, about whether this change is good or bad. The same chemical is a distinctive feature of many of the most expensive Sauvignon Blancs.

WHERE THEY'RE GROWN: Asparagus is an early-spring crop, with most harvesting finished by the end of April. California grows 80 percent of the fresh asparagus and more than half of all the asparagus in the United States. The next leading state is Michigan, which grows a little more than 10 percent of the total. In California the historic farming area is the deep, loamy delta of the Sacramento and San Joaquin rivers, around Stockton. There have also been significant plantings in the Imperial Valley, but in recent years these have dwindled due to competition from imported asparagus, mainly from Mexico and Peru.

HOW TO CHOOSE: Whether the asparagus is fat or thin, there are certain things to look for. First, check the tips. Remember that asparagus is a fern picked at an immature stage. The small "leaves" at the tip of the spear should still be tightly furled. The tender tip is also where asparagus will first begin to break down — it should be firm with no trace of softening. Second, check the base, which should be moist. Good markets sell asparagus sitting in a tray of water or on a moist towel to prevent the bases from drying out.

HOW TO STORE: Asparagus should be stored in the refrigerator in as humid an environment as possible. One way to accomplish this is to keep the spears upright in a container of water, like cut flowers. Drape a plastic bag over the top to create a moisture trap.

HOW TO PREPARE: To remove the tough bases, cut off the bottom inch or two of each spear: the thicker the spear, the more you'll want to remove. Thin asparagus can be prepared without peeling (though you may if you wish, of course). Thick asparagus must be peeled. Use a regular potato peeler and start trimming from the tip end with a light touch, gradually increasing the pressure as you get closer to the base.

ONE SIMPLE DISH: Steamed asparagus is one of the great pleasures of the spring table. Select the fattest spears you can find, then peel and trim them. Tie them together in a fat bundle with kitchen twine. Steam them until you can easily pierce one with the tip of a sharp knife. They should be quite soft, cooked enough that they sag slightly when you lift them by the twine, but not cooked so much that their color fades. Figure 7 to 8 minutes. When they are done, dress them with very good olive oil and fresh lemon juice, then sprinkle them with sea salt. Serve warm or at room temperature with lots of bread to soak up the juice.

Asparagus Wrapped in Crisp Prosciutto

This extremely simple recipe is a combination of two old favorites: basic roast asparagus and steamed asparagus wrapped in prosciutto. The contrast between the soft asparagus and the crisp prosciutto is delicious. Serve with iced champagne and plenty of napkins to wipe the asparagus juice from your fingers.

6 SERVINGS

> 1 **pound medium-thick asparagus (about 16 spears)**
> ½ **pound prosciutto, medium thinly sliced**
>   **(about 16 slices)**

Heat the oven to 450 degrees. Cut off the bottom 1 to 1½ inches of the asparagus spears. If the spears are thick, peel them. Wrap a slice of prosciutto around each spear spiraling upward, with the fatty stripe of the ham at the bottom so it creates a barber pole effect up the spear. Line a jelly-roll pan with aluminum foil and smear it lightly with olive oil.

Arrange the wrapped spears on the pan and place it in the oven. Do not disturb for 5 minutes, then shake the pan vigorously to turn the spears. Roast for another 5 minutes and shake vigorously. Continue roasting until the asparagus is very tender and the prosciutto is somewhat crisp, about 15 minutes total. Serve.

Asparagus and Shrimp Risotto

Who says homemade stock must be laborious? In this recipe, it's nothing more than quickly simmered trimmings from the main ingredient. Particularly when you're making a risotto from delicate ingredients, a "properly made" stock (meaning one you might use for a soup or sauce base) would be overpowering.

6 SERVINGS

1¼ **pounds thin asparagus**
1 **onion, minced (trimmings reserved)**
¼ **pound shell-on medium shrimp**
9 **cups water**
4 **tablespoons (½ stick) butter**
2 **cups risotto rice (preferably Arborio, Carnaroli or Vialone Nano)**
½ **cup dry white wine**
  Salt
3 **tablespoons freshly grated Parmigiano-Reggiano**
¼ **cup snipped fresh chives**

Cut off the bottom 1 to 1½ inches of the asparagus spears. Thinly slice the bottoms and add them to a large saucepan along with the trimmings from the onion. Shell the shrimp and add the shells to the saucepan. Cover with the water and bring to a boil. Reduce the heat to a bare simmer and cook for at least 30 minutes to make a mild stock before beginning the risotto.

Cut away the very tips of the asparagus and reserve them. Chop the shrimp into ½-inch pieces and add them to the reserved asparagus tips. Slice the remaining parts of the asparagus into ¼-inch rounds.

Place the asparagus rounds, 3 tablespoons of the butter and the minced onion in a large skillet over medium heat. Cook until the onion softens but does not turn color, about 5 minutes.

Add the rice and cook, stirring constantly, until all the kernels are opaque, about 5 minutes. Add the wine and stir until it evaporates. Ladle approximately 1½ cups of the simmering stock through a strainer into the rice and cook, stirring, until it evaporates. When the bottom of the pan is almost dry, add another ½ to ¾ cup stock and cook, stirring, until it evaporates.

Keep cooking this way, adding more stock as needed, until the rice begins to swell and become tender. Stir in 1½ teaspoons salt and the reserved chopped shrimp and asparagus tips. Continue cooking until the rice kernels are swollen and completely tender. Do not cook the mixture dry; the final texture should be somewhat soupy with slightly thickened liquid. This will take about 17 minutes in all.

Remove the skillet from the heat and add the remaining 1 tablespoon butter, the Parmigiano-Reggiano and the chives. Vigorously stir these into the risotto. The liquid will thicken even more. Taste and add more salt, if desired. Spoon into hot, shallow bowls and serve immediately.

MAKING RISOTTO

Much is made about making risotto, but essentially the dish is very simple: rice, main ingredient and stock. Add a flavoring base to start things out, a little wine in the middle and some butter and cheese to finish, and you've covered just about every possibility. But risotto's simplicity also makes demands.

You have to use the right rice — Arborio, Carnaroli and Vialone Nano are the ones most available in the United States — to get the distinctive creaminess. There are slight differences in the textures of these varieties, but not in taste. What makes them perfect for risotto is the way they are built. Like all rices, they are made primarily of starch: amylopectin and amylose. Amylopectin is a soft starch that dissolves readily in liquid; amylose is a hard starch that resists dissolving. Risotto rices contain far more amylopectin than most rices, although not as much as sticky Japanese sushi rice. When they are cooked properly, the soft starch leaks out and thickens the stock, forming what my friend the food and wine writer Matt Kramer calls "the sauce within." Generally, Arborio and Carnaroli, which are graded *superfino*, are a little higher in amylose (though certainly not nearly as high as long-grain rice). Therefore, they remain a little firmer, although they also tend to be a little stickier (particularly Arborio). Vialone Nano, which is *fino*, is a little softer and thickens the stock a little less, making it perfect for the slightly soupier style of risotto preferred in Venice. (Locals refer to that texture as *all'onda*, meaning it moves "like a wave.")

More important than which specific type of rice you use is the technique and the balance of ingredients. With risotto, the rice is the thing, and you don't want to overshadow it. One of the most common mistakes people make when fixing risotto is in the choice of liquid they add. Partly this is a misunderstanding of terms: Americans tend to use the words "broth" and "stock" interchangeably, but an Italian *brodo* is much lighter in flavor than what we normally think of. Make risotto with a

French-style stock, and stock is all you will taste. It would be better to use plain water. Although a good homemade stock will certainly make an exquisite risotto, you can do amazingly well with store-bought chicken broth thinned with water (do not use it straight). If you have trimmings from the main ingredient — say, asparagus peels, pea pods or shrimp shells — simmer them in the broth to add more flavor.

The making of risotto can be broken into four stages. An Italian cook I worked with gave them the names of the appropriate infinitive verbs, making a recitation sound like one of Dante's cantos: *soffriggere, tostare, bagnare, mantecare* (the four circles of risotto). This is much more impressive in Italian than in English. It really only describes four very basic operations: creating the flavor base (*soffriggere*, "to softly fry"), toasting the rice (*tostare*, "to toast"), cooking the rice (*bagnare*, "to bathe") and beating in the final addition of butter and cheese (*mantecare*, "to beat in fat" — what a joyous language to have a specific word for that).

We'll take one step at a time. All risottos start with some kind of flavor base, frequently nothing more than onions and butter or olive oil melted together over medium-low heat. You can add firm vegetables such as artichokes or the fat parts of asparagus at this point as well. Cook just until the ingredients start to shine. Do not let the onions brown.

Add the rice and stir to coat it with the flavoring base. Increase the heat to medium and keep stirring until you hear the rice "singing" as it scrapes against the bottom of the pan. This step sets the outer shell of the rice (mainly amylose) so that it will stay firm and won't get too mushy. You can actually see this happening: the outer perimeter of the kernel will turn translucent.

The first measure of liquid you add should almost always be wine. It doesn't take very much, but you do need to have some degree of tartness in the background (especially considering all that *mantecazione* you're going to be doing at the end). The wine will cook away in just a couple of minutes. Now begin adding the stock. The stock must be kept at a simmer the whole time. Adding cold stock to the rice or not keeping your risotto pan at a high enough temperature will delay the cooking and result in a gummy risotto. The first addition of the stock can be as much as 1 cup or even a little more. After you've given it a quick stir to distribute the rice, you don't really need to stir it again until the stock is almost gone. Add more stock when you can see a clean track in the bottom of the pan when you stir. With the second addition, reduce the amount of stock just a little. Again, once you've stirred the rice, you don't need to pay too much attention. When you can see a clean track in the bottom of the pan, it's time for the third addition. This is where you need to start paying attention. Add ½ to ¾ cup of stock and when that is almost gone, taste the rice. A properly made risotto will be slightly chewy rather than mushy, but there should be absolutely no crunchy uncooked starch left in the kernel. Right before you stir in that third addition is also the time to add any delicate, quickly cooked ingredients, such as asparagus tips, fresh herbs, shellfish or ingredients that may have been cooked in advance. When the rice is tender but still a little chewy, the risotto is almost done — but not quite. Add just enough stock to loosen the rice, then cook it just long enough to make it creamy.

Now you beat in the fat. This should be done off the heat to form a smooth emulsion that won't break apart. And it should be done vigorously — not only are you beating in the fat, but you're also bruising the rice kernels, squeezing out the last bit of starch and finishing the thickening.

Finally, risotto must be served immediately and in hot bowls. This is not a nicety but a necessity — a cold bowl or too long a wait will cool the risotto and set the starch, making the dish heavy and gummy.

Asparagus with Sauce Mimosa

Stirring together hard-cooked eggs, herbs and olive oil makes a sauce that is almost like a slightly chunky mayonnaise. Called sauce mimosa, it's one of those traditional preparations that seem to have largely disappeared — for no apparent reason.

6 SERVINGS

3 pounds fat asparagus
2 large hard-cooked eggs
2 tablespoons minced fresh parsley
1 tablespoon minced mixed fresh herbs (preferably chives, tarragon and chervil)
1 tablespoon fresh lemon juice
1 tablespoon sherry vinegar
3 tablespoons olive oil
   Salt

Bring a large pot of water to a boil. Cut off the bottom 1 to 1½ inches of the asparagus. Peel the asparagus, starting very lightly at the tip and gradually increasing pressure toward the base. Boil the asparagus just until a knife slips in easily, 4 to 5 minutes. Remove from the boiling water to a pan of ice water to stop the cooking.

Shortly before serving, peel the eggs and separate the whites and yolks. Chop them separately as finely as possible and then combine them in a small bowl. (Chopping them together smears the yolks.) Add all the herbs, lemon juice and vinegar and whisk together. Whisk in the oil. Add ¼ teaspoon salt, then taste and add more if needed.

Pat the asparagus dry and arrange on a platter. Spoon the sauce over the top across the middle of the spears. Serve at room temperature.

●●●● TO HARD-COOK EGGS

Cover the eggs with cold water in a small saucepan and set over high heat. When the water comes to a rolling boil, remove from the heat and let the eggs sit in the water until they are cool enough to handle.

Onions,
Leeks and Garlic

. . . .

Onions are so familiar that it seems impossible that there could be anything at all complicated about them. In fact, nothing is simple about this most basic of vegetables. Study them very long, and you may feel you've followed Alice down some fragrant rabbit hole. What seems to be one fairly consistent family turns out to be a wide and varied collection of many overlapping clans that parade under a bewildering assortment of names.

Onions can be red, white, yellow or green. They can be as big around as a softball or as tiny as a hazelnut. Some are even button-shaped. They can be called Spanish, Bermuda or Italian; spring, picklers, creamers, pearls or boilers. They can be sweet (those compose a virtual atlas — they can be from Maui, Walla Walla, Vidalia, Texas or the Imperial Valley, among other places). And then there are all those alliaceous cousins — garlic, shallots and leeks. Making sense of so many choices can be maddening.

Where to start? Probably the most important thing to know is whether they are sold fresh or dried. The vast majority of onions in the marketplace fit the latter category. They are grown to full maturity before harvesting; indeed, picking doesn't start until the bulbs have developed their full capacity of sugar and their green tops have withered and flopped to the side. A kind of chain dragged behind a tractor "undercuts" the onions, destroying the root systems and finishing the job. Usually, the fields are then mowed, removing the tops, and the bulbs are left to dry for more than a week before being harvested. Finally, the onions are removed to sheds, where fans circulate warm air constantly over the top. This curing process removes moisture, which can encourage fungi. Curing also sets the papery skin, which acts as a barrier against moisture and injury, further reducing the chances of spoilage. During this process anywhere from 3 to 10 percent of an onion's total weight is lost.

But onions also have a built-in mechanism for extending their shelf life, one with profound benefits for cooks. Most alliums are high in sulfurous compounds that naturally inhibit spoilage. Perhaps more important, however, these compounds give onions and their kin their distinctive character. Depending on the balance of specific components, the flavor can range from the sharp, grassy green of spring onions to the sweet mellowness of leeks to the orotund bass note of garlic. The base elements that form the flavors of most alliums (S-alkenyl derivatives of L-cysteine sulfoxide, if you really want to know) are constant. But the trace elements (1-propenyl-1, 2-propenyl-1, 1-propyl and methyl-L-cysteine sulfoxide) change, and those provide the distinctive differences.

What's really amazing is that in many cases those distinctive flavors — and even the chemical compounds that create them — are produced only when the flesh of an allium is bruised or cut. That damages the plant's cell walls, allowing the contents to combine. Enzymes react with existing compounds to create new ones, starting a split-second chain of chemical reactions that results in what we instantly recognize as onions' taste and smell.

The most common storage onions are the round brown ones (called yellow onions) that we often buy in big bags. They last the longest and tend to have the highest concentration of sulfur. White and red storage onions (both are sometimes called "Bermuda," which was once an important center of onion agriculture) tend to be slightly more delicate in flavor and don't store as well because they contain less sulfur. For this reason, they are frequently used raw. These are both round onions, but particularly in Italian neighborhoods, you can also find red onions shaped like torpedoes. These taste pretty much the same as round red and white onions. Small storage onions — called pearls, picklers, creamers, boilers and babies, depending on their size and the whims of marketers — come in all three colors as well. They are grown tightly packed together to limit their size. They all taste pretty much the same as well; their big draw is visual. The primary exception is the small, flattened Italian cipolla (also known by the diminutive "cipolline"), which has a fuller, rounder, sweeter flavor.

Other onions are picked fresh. The most common of these are the familiar pencil-shaped green onions. They are sold under a confusing variety of names — spring onions on the West Coast (though they are available year-round), scallions (on the East Coast) and, peculiarly, shallots (in Louisiana). Green onions are usually harvested well before they reach maturity and have begun to form a swelling bulb. Most important to a cook, they are picked before they have developed either much sulfur or much sugar. Their taste is mildly sharp and green (from the high chlorophyll content). Green onions also can be picked at the bulb stage, when the flavor is a little more developed. These are especially good for cooking whole, as the slightly larger size allows a pleasing range of textures.

To compound the confusion, there is also a branch of the family called "bunching onions." They look much the same as green onions, although their tops are a little more structured — when you slice them, they come away perfectly round rather than elliptical. The flavor is noticeably fuller and less pointed, as well as a little sweeter, than that of green onions. The king of the green onion is the Japanese variety called *negi* or *nebuka*, which is grown buried in the earth, like a slim, sharp leek. And then there are chives, which look like extremely fine green onions but are actually a separate species (*Allium schoenoprasum*). And garlic chives, which are still another species (*Allium tuberosum*). And that's not even addressing leeks (*Allium ampeloprasum*), which we'll get to later.

A third group of onions falls somewhere between storage and fresh onions. They are large and somewhat flattened and look like brown storage onions, but their shelf life is almost as short as that of the tenderest scallion. Indeed, these are the only onions that can truly be considered to have a season. So-called sweet onions are harvested and consumed in the spring, whether you're in Maui, Hawaii, or Vidalia, Georgia, or anywhere in between. These onions aren't so much sweet as they are low in sulfurous compounds (which is why they must be eaten right away and need to be refrigerated, whereas storage onions don't). They might more properly be called mild, not sweet, onions, and they are sometimes perilously close to being bland. In fact, whereas normal brown storage onions typically have a sugar concentration of about 12 percent, sweet onions average closer to 8 percent.

Perhaps because of their even lower concentration of harsh sulfurous compounds, well-grown sweet onions have an appeal that is almost fruity. They can be eaten raw, out of hand, just like an apple. In fact, they should be — cook them, and you only emphasize their lack of character. They are crisp, with just a little of the background burr that lets you know that you are, after all, eating an onion. And they cry out to be sliced for sandwiches or cut up for salads.

Sweet onions are a triumph of vegetable marketing. The first to achieve notoriety was grown in the early 1930s near Vidalia, Georgia, by a farmer named Moses Coleman. Vidalia was the local market center and was also the distribution center for a couple of major supermarket chains. Liking the local product, the supermarket managers began shipping it out to other member markets as early as the 1940s. By the 1970s more than 600 acres of sweet onions were being harvested in the area, and the farmers banded together to form a marketing cooperative. Because the Vidalia farmers were so successful, their neighbors in the nearby town of Glennville began promoting their own sweet onion. After a little intrastate sparring about who produced the sweetest onions (and facing even more competition from sweet onion growers from outside the state), the groups decided to unite and battle the world as one. Today more than 14,000 acres of so-called Vidalia onions are planted in a wide swath across the state of Georgia.

So popular are they that at least half a dozen other places market sweet onions of their own. In reality, not much differentiates those grown in Georgia from those grown in Maui, South Texas or almost anywhere else for that matter. They are almost all the same variety, usually called Grano or Granex, depending on the specific strain. This variety was developed in Texas from seed originally imported from the island of Tenerife off the coast of Spain. Scientists at a state-sponsored agricultural experimental station crossed this onion with another to come up with a whole string of sweet onions that they eventually exported to growing areas around the country. The sole exception is the sweet onion from Walla Walla, Washington, which has a somewhat tangled lineage. Walla Walla Sweets (it is a copyrighted trademark) descend from Italian seed acquired on the island of Corsica by a French soldier. He brought it to the United States when he immigrated in the early 1900s and introduced it to Walla Walla's Italian farmers.

Just because sweet onions are grown in so many places doesn't mean they can be grown anywhere. An onion's character is the result of a combination of genetics and environment. Some onion varieties tend to be more pungent than others. But a normally mild onion grown in the wrong place can pack something of a wallop or, alternatively, have no more flavor than water. Three environmental factors affect onion flavor. First is the chemical composition of the soil. Just as that distinctive onion flavor comes from sulfurous compounds, the strongest onions are grown in soil that is rich in sulfates. Second is watering. The strongest flavor belongs to onions that get the least water. Finally, temperature during the growing season can affect the pungency of an onion: the hotter the temperature, the hotter the onion. Most onions are planted in the spring and harvested in the fall, but sweet onions are planted in the fall in places where they can overwinter, to take advantage of the cooler temperatures.

As tangled as the whole onion family might be, that is just the beginning of the allium confusion. You also have to think about all those fragrant cousins — shallots, garlic and leeks.

Shallots look like brown pearl onions, except they are a bunching variety. This means that if you plant one shallot, you'll grow a whole clump of shallots, which will be joined at the base, with one side noticeably flattened where it grew next to its neighbor. This point is important to remember. There are onions that look quite a bit like shallots and are often sold as shallots. But they are not shallots, and the first clue will be that they lack that bottom junction scar and one flat side. Don't be fooled; they will also lack shallots' depth and complexity of flavor.

Garlic is so varied that it probably deserves its own family tree. Walk a farmers' market, and you might spot as many as half a dozen different varieties. Although each garlic has its own slightly different characteristics, there are two major families that you need to know about. The first type grows with a strong central stem, which remains after drying. This is called "hardneck" or rocambole garlic. It has the most varied and interesting flavors.

Through domestication, another family has been developed that lacks the strong central stem. It is called "softneck" garlic and is the dominant commercial type. Softneck garlic is easier to plant and grow, and it's certainly easier to pack and store. The trade-off is that the flavor isn't as good and clove size can vary tremendously within the same head. Elephant garlic is an entirely different species, close kin to the leek. Although it is large, its flavor is so mild that it's called "garlic for people who don't like garlic."

Leeks look like giant green onions. Unlike their smaller relatives, however, they are grown to full maturity. They retain their white bases by careful maintenance during the growing season. As soon as a little white becomes visible on the stem, dirt is piled around the leek. This shields it from sunlight, which in turn prevents the formation of chlorophyll (which would add a vegetal note to the flavor). This continuous piling on of dirt tends to make leeks extremely gritty; they must be washed thoroughly before you do anything with them. But any effort required is more than repaid by the results. Once leeks have been cooked, the flavor is deep and round, with a lingering sweetness. The texture is pure silk. This is one regal allium.

WHERE THEY'RE GROWN: California is the predominant state for onion farming, whether you're talking about storage onions (more than 25 percent of the U.S. crop), green onions (almost 50 percent) or garlic (more than 80 percent). Leeks and shallots are not tracked. Imports are also an important part of the picture. More than 10 percent of the storage onions consumed in the United

States come from outside the country — mostly Mexico, Canada and Peru. Americans also import more than 20 percent of all the garlic used, with China accounting for more than half of that.

HOW TO CHOOSE: Storage onions should be well filled, with the peel adhering to the bulb. There should be little shrinkage and no visible fungus (black dusty stuff). Avoid onions with green sprouts. Green onions should be fresh, with no decay at the tops and little shedding of the white peel. When green onions begin to get old, they are often trimmed, so check that the base is as wide as the root. When buying leeks, look for the same thing you do with green onions. Remember that the white part is what you eat, so the higher it goes on the stalk, the better off you are. Garlic should be well filled and solid. If the paper covers the cloves loosely, that's a sign of the shrinkage that comes with drying out. There should be no sprouting. Shallots should be well filled with a tight-fitting peel. Remember to check the base for the root scar and make sure one side is flattened to ensure that you're getting true shallots.

HOW TO STORE: Storage onions, shallots and garlic should be stored in a cool, dry place away from light. Sweet onions should be refrigerated, as should green onions and leeks.

HOW TO PREPARE: The easiest way to peel storage onions and shallots is to cut off the root and the top, leaving a strip of skin attached at each end. Pull down on that strip, and you'll create an opening in the paper from which you can easily peel the rest of the onion. Garlic can be broken out of the head by placing the head root end up on a cutting board. Press down firmly with the heel of your hand, and the cloves will pop free. To peel a clove, lay it on the cutting board and whack it just hard enough to bruise it with the flat of a chef's knife. That tight-fitting peel will come off easily. Because leeks are grown in mounds of dirt, they require special care in cleaning. Cut off the roots, leaving the base intact. Cut off the tough green leaves (these can be frozen to use in stock). Leaving about 2 inches of the base intact, cut the leek lengthwise into quarters and rinse it well under running water. Inspect it carefully for any hidden grit.

▶ ONE SIMPLE DISH: Caramelized onions are a preparation that belongs in every cook's repertoire. They are easy to make and endlessly versatile: add a little cooked bacon or slivered prosciutto, and you have a pasta sauce; top a pizza with caramelized onions and goat cheese, and you have a feast. Caramelizing onions is basically a matter of long, slow cooking, but there are a couple of tricks that make it easier. Slice the onions into rings that are ¼ to ½ inch thick, depending on how long you want to cook them and how much texture you want at the end (thicker equals longer with slightly more bite). Put them in a cold, heavy skillet with just enough olive oil to lightly cover the bottom of the pan. Start them cooking over medium heat. When the onions have softened a little, salt them, cover the pan and reduce the heat to low. The salting will draw moisture from the onions; the lid will concentrate the heat. Cook them, stirring every 10 to 15 minutes, until limp and beginning to color, about 45 minutes. Raise the heat slightly and remove the lid. They will finish coloring in another 5 minutes. If you want to add garlic (something I heartily recommend), do it only at the last stage. Garlic contains much less moisture and will scorch a lot earlier.

Pink Pickled Onions

This is one of those preparations that in an ideal world would always be in our refrigerators. The sharpness of the vinegar and the sweetness of the onion combine to make a perfect addition to almost any sandwich you can think of. These onions also are good on sliced roast meat, particularly pork.

MAKES ABOUT 1 QUART

2 cups water
2 cups white vinegar
4 whole cloves
1 bay leaf
½ teaspoon dried oregano
1 red onion, sliced into ½-inch rounds

Bring the water, vinegar, cloves, bay leaf and oregano to a boil in a medium nonreactive saucepan and remove from the heat. Let steep until cool, about 1 hour.

Place the onion rings in a sealable container and pour the vinegar mixture over the top. Cover tightly and refrigerate for at least 2 hours. The pickle will improve greatly if left for 24 hours and then will level off. The onions will be good for at least a week.

Grilled Cheese Sandwiches
with Sweet Onions

This is a kind of dressy turn on everybody's favorite, the grilled cheese sandwich. Prepared this way, these sandwiches are good enough to serve as a passed appetizer at the fanciest dinner party, and they're unbelievably good with champagne.

8 SERVINGS

1 loaf firm, finely grained white bread, such as a pullman loaf, unsliced
1 pound semisoft cheese, such as Brie, Taleggio or Teleme
1 large sweet onion (about ¾ pound)
1 teaspoon champagne vinegar
2 teaspoons minced fresh parsley
Salt
Butter

Trim the crusts from the loaf of bread, leaving a solid, evenly shaped rectangle. Using a serrated knife, carefully cut the bread lengthwise into slices just as thin as you can, ¼ to ½ inch thick. You should get 6 to 8 slices.

Trim the rinds from the cheese and cut it into thin slices. Cut the onion in half through the stem end and then slice it crosswise as thinly as possible, using a mandoline or a very sharp chef's knife. Put the onion in a bowl and season it with the vinegar, parsley and salt to taste.

Arrange one fourth to one third of the cheese over 1 long slice of bread, being careful not to come too close to the edges. (You don't want the cheese oozing into the pan when it melts.) Arrange one fourth to one third of the onion over the cheese. Cover with another slice of bread and press together firmly. Repeat with the remaining bread, cheese and onion.

Heat 1 tablespoon butter in a large skillet over medium-low heat. Fry 1 sandwich at a time, gently but firmly pressing down with a spatula or heavy pan. Cook until well browned, about 6 minutes on the first side, about 4 on the second. Repeat with the remaining sandwiches, adding more butter as necessary.

Neatly trim the edges of the cooked sandwiches with a serrated knife, then cut each sandwich into crosswise strips about ¾ inch thick. Serve hot.

Sweet Onion, Avocado and Shrimp Salad

Sweet onions have become a bit of a cliché, but they certainly have their charms when used right. This is one dish where they work well. By giving the onions a quick soak but otherwise not cooking them, you moderate their already mild bite while still preserving every bit of their crispness.

6 SERVINGS

  1 **large sweet onion (about ¾ pound)**
1½ **pounds avocados (2–3)**
    Fresh lemon juice
  ¼ **cup olive oil**
    Salt
  2 **tablespoons snipped fresh chives,
    plus more for garnish**
  1 **pound medium shrimp, cooked and peeled**

Cut the onion lengthwise into quarters, then cut each quarter lengthwise into halves or thirds. Cut each of these sections in half crosswise. Place the onion pieces in a bowl of cold water to soak for at least 10 minutes.

Cut the avocados in half, remove the pits and with a spoon scoop out the flesh in one piece. Cut the flesh into pieces roughly the same size as the onions.

In a large bowl, whisk together 2 tablespoons of lemon juice, the olive oil, salt and chives. Drain the onions and pat them dry with paper towels. Add the onions to the dressing and stir to coat well. Remove the onions with a slotted spoon, draining any excess dressing back into the bowl, and arrange them on a platter in a broad layer.

Add the avocados to the leftover dressing and stir to coat well. Remove them with a slotted spoon and arrange them in an oblong mound on top of the onions, centering the mound so the onions show around the edge.

Add the shrimp to the leftover dressing. Add a squirt of lemon juice and a little more salt to taste. Stir to coat well. Arrange the shrimp on top of the avocados and garnish with snipped chives. Serve.

PLATTER SALADS

"Composed salad": the very name is enough to kill your appetite, evoking visions of white-gloved matrons sitting around after bridge eating canned sliced peaches artfully arranged over slivered iceberg lettuce and decorated with cream cheese rosettes. In the minds of most people, composed salads are antiques from the old-fashioned school of cooking that was concerned more with the way dishes looked than how they tasted. After all, why arrange all those ingredients so carefully if you're just going to toss them together at the table?

But let's not be too hasty to poke fun. There's nothing wrong with composed salads that a bit of updating won't cure. Like much of classical cuisine, when composed salads are stripped to their barest components, there is good, sensible food behind the stilted prettiness. In fact, among better cooks, the decoration has never been the dish's main virtue. In that bible of old-fashioned French cooking, the *Larousse Gastronomique*, the term *salades composées* is translated as "combination salads" (as opposed to "simple salads"), emphasizing the mix of raw and cooked ingredients rather than their artistic arrangement. Escoffier went one step further, speaking out forcefully against the overdecoration of the dish. "The increased appetizing look resulting therefrom is small compared with the loss in the taste of the preparation," he wrote in *The Escoffier Cookbook*. "The simplest form of serving is the best, and fancifulness should not be indulged in."

So forget about separating the meat, vegetables and greens into little decorative piles. Arrange them in a more modern, naturalistic way, and you have something delicious that is also beautiful without being contrived. Even better, you have dinner. Because when you get right down to it, a composed salad is the perfect meal for a hot summer night. Take a small portion of fish, meat or cheese and arrange it on a colorful bed of vegetables, greens and herbs. Bind the whole thing together with a bold dressing of some sort. What could be better?

Actually, you're probably already making composed salads right now without even realizing it. Ever slice ripe tomatoes and dripping-fresh mozzarella, decorate them with dark green fresh basil and then serve the dish with good olive oil and a loaf of crusty bread? Is there a better — or prettier — dinner on a sweltering weeknight? That's a very basic example. How about tossing together canned white beans and tuna, a little olive oil and lemon juice and some sharp bites of chopped red onion? Or what about thinly sliced steak and room-temperature steamed potatoes, bound together with a mustardy vinaigrette?

Whereas "composed" may once have referred to how the salad was arranged (preferably in as static and staid a way as possible), now it has more to do with flavor and the interplay of taste and texture. Although presentation is still important, the fashion today is for salads that look like food rather than a painstakingly arranged still life.

What may be even more impressive than appearance is the way these salads adapt to what you already have in the pantry and refrigerator. Canned white or garbanzo beans, tuna and smoked salmon; leftover grilled chicken or steak and the vegetables from a big Sunday dinner. All you need is some sturdy greens, a good dressing and a few condiments, and you're in business.

Since these salads are served at room temperature, it's best to use meat that is fairly lean. So remove the skin from the chicken. Or if you're using beef, flank steak is a good choice. Lately, with more and more groceries stocking meat cut as it is in Mexico, you can also find very thinly sliced sheets of skirt steak — the kind that's usually used for *carne asada*.

Through the summer, try to keep at least one head of watercress, frisée, radicchio or curly endive in the crisper. These greens have the strength — in both structure and flavor — to match up to almost anything you can throw in the bowl.

Make sure the meat and all the vegetables are cut into bite-size pieces, preferably of a similar size. These salads are about balanced combinations of flavors; don't let any one ingredient dominate.

The dressing is the unifying factor in a composed salad. For lighter combinations, such as tuna and garbanzo beans, a simple mixture of olive oil and lemon juice is all that's needed to point up the flavors. As the ingredients get heavier and bolder, so should the dressing. Adding a little Dijon mustard and some minced shallots to that vinaigrette will make a nice accompaniment to a salad made with grilled beef. Flavored mayonnaise is always good. Start with either homemade or a good commercial brand (Best Foods, sold as Hellmann's on the East Coast, is the perennial first choice). Then add minced herbs, pureed roasted peppers or whatever else strikes your fancy. Even the best of the prepared mayonnaises will benefit from the addition of a little acidity: lemon juice or a vinegar of some sort.

Whichever dressing you choose, use just enough to season the salad and give it cohesiveness, but don't overdo it. When you taste the salad, the main flavor should be the ingredients, not the dressing. And a good salad should never be gloppy. A rule of thumb is to add half the dressing and toss well, then add the rest a tablespoon at a time. Remember, you can always pass more dressing at the table.

Also remember that if you're using starchy ingredients — beans, rice or potatoes, for example — they'll absorb the flavors better if you mix them with a little bit of dressing while they're still hot.

Peas and
Fava Beans

. . . .

Among mankind's strongest cravings are those for sweet-ness and ease. That explains why springtime cooking has always been such a paradox. To get sweetness, you have to work for it. There is no better example than legumes.

Just sorting them all out is hard enough. The word "legume" refers specifically to plants that carry their seeds sealed in pods. Since English peas and favas beans are so similar, for the sake of culinary convenience, we'll lump the two together. Most people are familiar with English peas, but fava beans are a relatively new addition to the American plate. Favas are much larger than peas, both before shelling and after. In the husk, they look like gigantic, slightly fuzzy English peas. The actual bean is flattened, looking more like a lima bean than a round pea, and depending on its ma-turity, after shelling it can range in size from about the same as a pea to as big around as a nickel.

English peas and favas appear in the same season. Both are in-comparably sweet, green and delicious — as well as exasperatingly labor-intensive. Peas need to be shucked of their pods — a chore that, given a communal setting with friends sitting around the kitchen, can even seem like rustic fun, like a quilting bee or some-thing. Sylvia Thompson, in *The Kitchen Garden,* noted that it takes 25 pea plants to produce 1¼ cups of shelled fresh peas — barely a serving for a real pea fan. Fava beans are almost as bad. You have to clean 3 pounds of fava pods to get less than 2 cups of beans. Favas need to be shucked and then, a single bean at a time, peeled of their tough skins. There's no way to put a pretty face on it: this is more like peasant drudgery. But just when you're sitting there, elbow-deep in bean pods, thinking there is no way you'll ever go through this again, you pop a bean in your mouth and are rewarded with an almost lightning explosion of sweet green flavor that somehow seems to sum up the entire beauty and promise of spring in a single burst.

That peas have come to represent spring is not just some accident of gustatory symbolism. Agriculturally, they are uniquely suited to the season. In the first place, they thrive in much cooler weather than most vegetables. English peas grow best when the temperature remains below 65 degrees. A heat spike up to 80 degrees can kill an entire field. Also, peas mature much more quickly than most other vegetables. A pea plant must be started from seed, but it can be ready to harvest only forty-five days after planting. Broccoli, another cool-weather crop, can take as much as three times that long. Favas take three or four weeks longer than English peas (about eighty-five days normally). This is one reason they are more popular in the Mediterranean, where they can be planted in late fall and will survive the mild winter.

Legumes have the happy trait of nurturing the ground in which they're grown as well as the humans who grow them. The roots of legumes contain nodules that host bacteria called rhizobia. These bacteria have the ability to convert nitrogen from the atmosphere into soilborne forms that can be used by plants, reducing the need for fertilizer. Plant legumes for the spring, and the field will be even more fertile in the fall.

The bright, sweet flavor is the main attraction of peas and beans, but it is as fragile as it is appealing. How do you maintain it from field to plate? Speed and refrigeration are the answer. The sugar that gives fresh peas their flavor vanishes quickly, particularly if they are not stored carefully. Let peas get warm, and that sugar will convert very quickly to starch. This happens almost ten times faster at 70 degrees than it does at 32 degrees. (The latter is perfect for storage, but it is only one degree warmer than the freezing temperature for peas — scant margin in a home refrigerator.) Even storage temperatures warmer than 45 degrees — about the temperature in the top of your refrigerator — can lead to toughening and rapid yellowing. Since storing peas is obviously such a problem, put them in your "dinner the same day" category.

Although English peas and fava beans share so many traits both delicious and exasperating, there is one big difference between them, and that is the maturity at which they should be picked. Peas are best when they are fully mature, so that the little seeds have swollen in the pod. (*Petits pois,* or baby peas, are not immature peas, but a separate variety that is naturally smaller.)

Fresh favas are best when they are truly babies. Tiny fresh favas are all the rage in Rome during the spring, where they are picked straight from the pod and eaten with moist springtime pecorino Romano, a salty sheep's milk cheese that perfectly offsets their sweetness. (Feta would be a good alternative.) Favas of that size are so small that the peas don't bulge in the pod. As long as the peas are less than ½ inch in diameter, favas can be cooked without peeling their outer skins. Another indicator of freshness is the color of the skin: when favas are young, the skin is green; as the bean matures and toughens, the skin turns white. Unlike peas, favas are useful even when they are overmature and starchy. Cook large favas until they are soft and then puree them with a knob of butter.

Sugar snap and snow peas are alternatives to delicate English peas. They are classed by the ag folks with the unpoetic though descriptive tag "edible pod peas." These are both venerable varieties, in which the pod is as sweet as the pea (or nearly so, anyway). Both fell out of favor in the United States at the turn of the twentieth century, only to be reintroduced in the foodie 1970s — snow peas by a thousand Chinese restaurants, and sugar snaps by the enthusiastic marketing of a plant breeder named Calvin Lamborn, who had developed an improved variety.

Lamborn had the bright idea to cross a "sport" pea plant that had formed very hard, tightly sealed walls with the snow pea. His version, introduced in 1979, was so successful that the name "sugar snap," which originally applied only to his specific variety, has now become acceptable for generic use. Previously, these peas had been called "sugar peas" or "butter peas."

Snow peas have flat pods; sugar snaps are rounded. Sugar snaps tend to be exuberant: crunchy and very sweet. Snow peas are somewhat subtler. Both will almost always be sweeter than all but the best English peas. This is not because they withstand storage better, but because they start out with more sugar and so are better able to afford the inevitable degradation that storage brings. Although they may be sweeter, neither variety captures the full green "pea" flavor of English peas or favas. Both should be cooked only briefly, if at all. More than a minute or two of blanching or stir-frying, and the advantages are lost.

English Peas and Favas

· · · · · · · · · · · · · · · · · ·

WHERE THEY'RE GROWN: Neither fresh English peas nor fava beans are grown widely enough to be tracked statistically. English peas are grown for drying, canning and freezing in several states, but predominantly Minnesota, Washington and Oregon. They are also imported from Mexico, the Caribbean and Central America.

HOW TO CHOOSE: Pods should be firm, green and unwrinkled. A little blackening sometimes occurs in fava bean pods. English peas should be swollen in their pods. It is generally considered acceptable to pop a pod and taste a pea before buying. Fava beans should show only as small bumps.

HOW TO STORE: English peas should be stored as briefly as possible, tightly wrapped and refrigerated at the coldest level (in the crisper drawer). Fava beans are more forgiving.

HOW TO PREPARE: English peas should be shucked straight into a pan for cooking — but do not shuck them in advance. Fava beans take an extra step because of their skins. The simplest way to remove the skins is to shuck the beans into a large bowl, then pour boiling water over the top. Let them stand for several minutes before draining. You should be able to break the tight, lighter-colored inner skin with your thumbnail and then squeeze the bean out between your fingers.

ONE SIMPLE DISH: Both English peas and fava beans are delicious simply braised with a little heavy cream and ham. Cut the ham into tiny cubes and render it in butter. Add some minced shallots and then some cream. When the cream has reduced slightly, add the peas or favas and cook just until they are tender and sweet.

Snow Peas and Sugar Snaps

WHERE THEY'RE GROWN: Like English peas and favas, neither is grown widely enough to be tracked statistically. They are also imported from Mexico, the Caribbean and Central America.

HOW TO CHOOSE: Pods should be firm and crisp. Reject any that show signs of wilting. Sugar snap pea pods may show traces of scarring, which does not affect the flavor.

HOW TO STORE: Store as briefly as possible, tightly wrapped and refrigerated at the coldest level (in the crisper drawer).

HOW TO PREPARE: Sugar snap peas and snow peas still have tough fibrous strings that run the length of the pods and must be removed before cooking. Check carefully: some varieties have strings on both sides of the pods.

ONE SIMPLE DISH: The less you cook a snow pea or sugar snap, the better off you'll be. Anything more than a lightning-quick blanching or stir-frying destroys the crisp texture and bright flavor, which are their most delicious qualities.

Sugar Snap Pea Soup with Parmesan Cream

This soup can also be served cold, with a few fresh chervil leaves rather than the Parmesan cream. And it makes a nice small appetizer if presented in espresso cups, in which case it will serve at least 12.

6 SERVINGS

- 2 pounds sugar snap peas, strings removed
- ¼ cup minced shallots
- 2 tablespoons butter
- ½–¾ cup chicken broth
- Salt
- Freshly grated nutmeg
- 1½ teaspoons fresh lemon juice, plus more to taste
- Up to 1 teaspoon sugar
- ½ cup heavy cream
- ¼ cup freshly grated Parmigiano-Reggiano

In a large pot of rapidly boiling, generously salted water, cook the sugar snap peas until they are quite tender but still a vibrant green, 6 to 7 minutes. Do not cook so long that they turn drab. As soon as the peas are done, drain them and place them in an ice water bath to stop the cooking and preserve their bright color.

While the peas are cooking, cook the shallots in 1 tablespoon of the butter in a small skillet over medium-low heat until tender and translucent, about 5 minutes. Set aside.

Place half of the peas in a blender and puree until very smooth. Add a tablespoon or two of chicken broth, if necessary, to keep the mixture flowing. Add the remaining peas and the cooked shallots and finish pureeing.

Pass the pea puree through a strainer into a bowl, pressing with the flat blade of a rubber spatula to work it all through. Rinse the spatula blade to remove any fiber and scrape the thick pea puree that sticks to the outside of the strainer into the bowl. Discard the fiber that is left behind inside the strainer.

Stir just enough chicken broth into the puree to make it a flowing liquid. It should have the consistency of fairly thin split pea soup. Stir in 1 teaspoon salt, a few gratings of nutmeg and the lemon juice. Taste, and if the peas aren't bright and sweet, stir in enough sugar to correct. If necessary, add more salt and lemon juice as well. (The recipe can be prepared to this point up to 8 hours in advance — any longer, and the color will start to fade. Refrigerate in a tightly covered container.)

Pass the puree through the finest strainer you have into a saucepan. Warm over medium-low heat until the mixture is bubbling. Stir in the remaining 1 tablespoon butter. Taste and adjust the seasonings once more.

While the puree is warming, cook the cream and Parmigiano-Reggiano in a small saucepan over medium heat just until the cheese melts and the cream is thick enough to coat the back of a spoon.

When the puree is hot, divide it evenly among six soup plates. Shake each plate gently to distribute the puree in an even layer. Spoon some of the Parmesan cream into the center of the puree in a rough C pattern. Serve immediately.

Sugar Snap Peas and Shrimp
with Chive Mayonnaise

Here is a remarkable dish that comes together in minutes yet still looks and tastes so elegant it's appropriate for serving at the fanciest dinner party you can imagine.

6 SERVINGS

- 1 **pound shelled medium shrimp**
- 1 **pound sugar snap peas, strings removed**
- ⅓ **cup mayonnaise**
- 2 **tablespoons snipped fresh chives**
- 1 **teaspoon minced shallots**
- 1 **teaspoon fresh lemon juice**
  Sherry vinegar

Cook the shrimp in plenty of rapidly boiling salted water just until firm, about 5 minutes. Drain and chill in ice water to stop the cooking. Pat dry and combine with the sugar snap peas in a large bowl.

In a small bowl, whip together the mayonnaise, chives, shallots and lemon juice. Beat in a splash of sherry vinegar, then taste and add more if necessary.

Spoon two thirds of the mayonnaise over the shrimp and sugar snaps and stir together. The mayonnaise should lightly and uniformly coat both. If necessary, stir in a little more mayonnaise. Refrigerate before serving.

Crisp-Skinned Salmon with Braised Spring Peas and Mushrooms

I learned this trick for preparing salmon from the French Laundry chef Thomas Keller. This dish is also delicious made with a pound of fresh pea sprouts instead of the English peas. Just toss them in at the end and cook only until they begin to wilt.

6 SERVINGS

- 2 pounds center-cut salmon fillet, in one piece
  Salt
- 2 slices bacon
- ½ pound crimini mushrooms
- 3 shallots, minced
- 6 sprigs thyme
- ½ cup dry white wine
- ¼ cup heavy cream
- 1 pound shelled English peas (fresh or frozen)
- 1–2 tablespoons vegetable oil
- 1 tablespoon butter

Place the salmon skin side up on a cutting board. Run the back of a knife over the skin, using a squeegee motion. As moisture emerges from the skin, wipe the skin and the knife dry with a paper towel. Repeat until no more moisture is visible.

Turn the salmon over and feel along the surface of the flesh with your fingertips just above and below the midline. If you feel pin bones, pluck them out with tweezers or needle-nose pliers. Slice the fillet in half lengthwise and then into thirds crosswise to make 6 somewhat square fillets. Season the meat side with salt and set aside skin side up until ready to cook.

Slice the bacon into thin crosswise strips. Cook them in a large skillet over medium heat until they are slightly crisp and the fat has rendered, about 10 minutes. Remove the bacon with a slotted spoon and reserve, leaving the fat in the pan.

While the bacon is cooking, trim the dried-out base of each mushroom stem and cut the mushrooms lengthwise into quarters. Cook the mushrooms in the rendered bacon fat until they color slightly, 2 to 3 minutes. Add the minced shallots and cook until they soften but do not color, 2 to 3 minutes.

Return the bacon strips to the pan and add the thyme and white wine. Cook until the wine has reduced to a thin glaze, about 5 minutes. Add the cream and peas and cook just until the peas soften. This will take 5 to 10 minutes for fresh peas, depending on their starchiness, and about 5 minutes for frozen. Remove from the heat until ready to serve.

Heat a large nonstick skillet over high heat and add just enough oil to film the bottom of the pan. When the oil is hot enough to sizzle, place the salmon squares skin side down in the pan and reduce the heat to medium. Cook until you see the cooked color come about one third of the way up the side of the fillet, 3 to 5 minutes. Remove the pan from the heat and turn the salmon over to finish cooking off the heat.

Place the peas and mushrooms over high heat and stir in the butter. Season with salt to taste. Divide the mixture among six shallow pasta bowls and place a salmon square skin side up in the center of each. Serve.

Risotto of Fava Beans, Baby Artichokes and Spring Onions

Fava beans are a pain to peel, but how can you celebrate spring without them? Make a few of them go a long way by combining them in a risotto with baby artichokes (another spring favorite). Shaving the cheese with a vegetable peeler gives you delicate sheets that soften over the hot risotto.

6 SERVINGS

- 1 pound fava beans, in the shell
  Juice of 1 lemon
- 1½ pounds baby artichokes
- 1 bunch spring onions or green onions
- 1 14½-ounce can chicken or vegetable broth
- 2 tablespoons olive oil
  Salt
- 3 cups Arborio rice
- 2 tablespoons butter
- 4 sprigs mint
- 1 ounce pecorino Romano

Shell the favas and place them in a large bowl. Cover with boiling water and let stand for 5 minutes. Drain the beans in a colander and rinse in cold water. Peel the tough skins by cutting one end with your thumbnail and squeezing the bean out with your thumb and forefinger. The bean will shoot out; aim carefully into the bowl. Set aside.

Fill a bowl with cold water and add the lemon juice. Remove the tough outer leaves of the artichokes until you come to pale green leaves. Cut away the top third of the leaves. Using a paring knife, peel the tough green skin from the base and stem of each artichoke. (Do not remove the stems.) Cut lengthwise into quarters. Set aside in the lemon juice and water.

Trim the roots and dark green tops from the onions, then cut them lengthwise into ½-inch-thick segments.

In a large pan, combine the broth and enough water to make 8 cups and bring to a boil over high heat. (You will probably need only 6 to 7 cups, but the broth will reduce during boiling.)

Heat the olive oil in a large skillet over medium-high heat. Remove the artichokes from the water and add them to the skillet along with the onions. Cook, stirring occasionally, until the artichokes and onions begin to soften, about 5 minutes. Season with salt to taste.

Add the rice and cook, stirring, until all the grains are coated with oil and the outer covering becomes slightly translucent, about 3 minutes.

Begin adding the hot broth, starting with 1 cup at a time. Cook, stirring, until the broth has all but evaporated, leaving just a film in the bottom of the pan, about 5 minutes. Add another cup of broth and repeat. After the second cup, begin adding broth ½ cup at a time, repeating the procedure until the grains are swollen and tender to the bite. The rice should be firm but not chalky at the center.

Remove from the heat and immediately add the fava beans, butter and 4 or 5 mint leaves. Stir vigorously to incorporate the butter. Season with salt to taste and serve immediately, garnishing with the remaining mint leaves and shaving sheets of pecorino Romano over the top with a vegetable peeler.

Salad Greens

• • • • • • • • • • • • •

It's easy to poke fun at iceberg lettuce. About the best thing you can say for it is that it's crisp and it doesn't taste bad. Of course, it doesn't taste particularly good either. In almost every way you look at it, iceberg lettuce is nothing more than crunchy water.

But when viewed from a certain angle — that of industrial engineering — iceberg lettuce becomes a lot more interesting. An iceberg lettuce field is about as close to a widget factory as farming can get. Iceberg lettuce is grown on a massive scale at an amazingly constant rate (seasons mean little or nothing to the iceberg industry). It is firm enough that it can be picked and packed with a minimum of trouble, and it is hardy enough that it can be stored as long as three weeks without noticeable ill effect. In short, it is just about the perfect product for industrial agriculture. But like so many other industries, the growing of iceberg lettuce is changing rapidly.

Lettuces were almost certainly among the first plants humans ate. After all, in their most ancient form, they were nothing but wild leaves. They also were among the first crops domesticated. Egyptian art from about 2500 B.C. depicts fairly accurately a variety of lettuce that is quite similar to one that is still being grown in Egypt today. This is not a lettuce most of us would recognize. It resembles romaine, but its leaves grow out of thick stems that look like celery stalks.

For the second half of the twentieth century, however, if you said "lettuce," it was pretty much assumed you meant one thing: iceberg. Introduced in the 1940s, iceberg lettuce is a refinement of an old lettuce family generically called "crisphead." In Europe these and similar varieties are called Batavias. In the 1920s, in response to a lettuce blight that was threatening the industry, a USDA plant breeder named I. C. Jagger began a series of refinements that resulted in the first true iceberg lettuce, called Great Lakes, which was released in 1941. (A named variety called Iceberg predates this lettuce, but that is coincidental; it is softer and smaller than true icebergs.) Descendants of the Great Lakes family, called the Salinas family, still dominate the iceberg industry. There are dozens of strains, each specifically developed for a certain climate, growing area or production window.

Iceberg is a very good lettuce, if crunch is all you care about. For more than forty years, its ease of growing and resistance to disease and rough handling made it by far the dominant lettuce in the country. A leaf of iceberg inevitably went on nearly every hamburger sold; shredded, it topped almost every taco.

Iceberg is grown up and down the state of California and even into the Arizona desert, depending on the time of year. (California alone produces almost three quarters of the iceberg grown in the United States, and the two states combined grow about 98 percent.) Iceberg is a cool-weather crop, growing best when daytime temperatures are in the low 70s, cooling to the mid-40s at night. If the daytime weather gets into the 80s or nighttime lows are in the 50s, the lettuce can bolt, sending up shoots with flowers and seeds and turning bitter. To catch those optimal temperatures through the different seasons, the iceberg harvest is almost constantly on the move. During the winter, growers work ground in the desert areas along the Colorado River in California's Imperial Valley and around Yuma, Arizona. As the weather warms, they move to the Central Valley for a couple of months. Iceberg is also grown along California's southern coast, from northern San Diego to Santa Barbara.

But just about the perfect location for growing lettuce is the Salinas Valley, about one third of the way from San Francisco to Los Angeles. Here the pressure cooker heat of the Central Valley is moderated by a break in the Coast Ranges leading to Monterey and the Pacific Ocean. The Salinas Valley calls itself the Salad Bowl of the World, and that description is only slightly exaggerated. The first major planting of lettuce went into the Salinas Valley in the early 1920s, and lettuce acreage quickly skyrocketed. In 1922 roughly 300 acres were planted with iceberg; by the end of the decade, that number had risen to 43,000. Today there are about 65,000 acres of iceberg in the valley — a little less than half the state's total. Indeed, driving through the valley during most of the year, you spend what seems like hours passing through lettuce field after lettuce field with little else in between. That's because not only is lettuce grown on a massive scale in the Salinas Valley it is also grown for a long time.

Whereas other areas can harvest for only two or three months before the weather turns too hot or too cold, in the Salinas Valley lettuce can be cut from April through November. Between planting and harvesting, lettuce is nearly a year-round crop there. In fact, there is a county-enforced break in lettuce production during the month of December to break the life cycle of certain viruses. Farmers there traditionally spend the long Thanksgiving weekend breaking down the equipment necessary for harvesting and packing lettuce, then trucking that equipment south to Yuma or the Imperial Valley.

Although lettuce is grown on an industrial scale, it is still sold like a commodity. A case of iceberg fluctuates in value on a daily basis. The weird thing is that for much of the year, the farmer usually gets paid less for the lettuce than it costs him to grow, harvest and ship it (on average, approximately $10 per twenty-four-head carton).

If this sounds like a questionable business strategy, welcome to the world of iceberg economics. Lettuce farmers are agriculture's version of day traders or riverboat gamblers. They break even or take losses many weeks in hopes that somewhere down the line, they'll catch weeks when everyone else runs short of lettuce and the price takes off. This is one reason the same grower will usually maintain fields in several different parts of the state. Because so much iceberg lettuce goes to fast-food restaurants, the demand is fairly constant. All it takes for a price bump is a slight hiccup in supply. This happens regularly but unpredictably. A spell of unseasonable weather can double prices overnight.

And every once in a while, the price does more than bump up. In March 1995 more than 10,000 acres of farmland around Salinas flooded, wiping out as much as half of the lettuce supply. The price of a case of iceberg, normally around $10 to $12 in March, doubled to $24. Then it went to $36. Then to $60. There were reports of lettuce prices of up to $100 a case. How expensive was iceberg lettuce? Some restaurants started serving fancy arugula salads at staff meals because even that was cheaper. Of course, it doesn't take an event of that magnitude to cause prices to rise. A couple of months later, after the harvest had recovered and prices had returned to normal, a case of iceberg went back up to more than $25 when a brief hot spell caused light picking.

Still, the glory days of iceberg farming are probably in the past. In the mid-1970s iceberg accounted for more than 95 percent of the lettuce production in the United States. But then tastes changed, and so did iceberg's fortunes. This happened in two different directions. At the high end came the advent of the Chez Panisse salad, a mix of various kinds of tender leafy greens that were almost unknown at that time, all tossed together to produce an interesting combination of flavors and textures.

At roughly the same time, what could be called the "Caesaring" of the American salad began. Restaurants as varied as Burger King and the French Laundry began selling some variation of the Caesar salad, which is traditionally made with romaine. By 2000 iceberg's share of the lettuce market had plummeted to less than 60 percent, while romaine's had climbed from nearly nothing to roughly 20 percent. In 1993 about 16,000 acres of romaine were grown in California; ten years later there were more than 60,000.

But even those statistics somewhat underestimate iceberg's fall from grace. Because during the 1990s, there began another revolution in the lettuce fields — bagged salads. Rather than buying whole heads of lettuce and mixing them at home, new technology allowed cooks to buy greens that had already been cut and mixed at the packing shed. At first this product was intended mainly for institutional use — big bags of chopped salad would save lots of labor in hotel and hospital kitchens. But today between 35 and 40 percent of the lettuce grown in the United States winds up in one of those salad bags.

Although rarely trumpeted on the label, iceberg, shredded into pieces, is often included in bagged mixes. Americans, it turns out, still want to have some crunch, even in supposedly sophisticated salads. But if you take away the iceberg that goes into the shredder you'll find that the production of whole-head iceberg has fallen to less than 40 percent of the lettuce crop today. Ironically, while iceberg lettuce is taking a dive here, it is being increasingly planted in Europe, where it is perceived as a hot new thing.

Leading the charge against iceberg is romaine, which is called cos in Europe. Romaine is an entirely respectable lettuce, combining what is good about iceberg (the crunch) with a sweet, more emphatically green flavor. The other main families of lettuce grown in the United States are leaf, which forms very loose, rosette-shaped heads (think green and red oak leaf), and butterhead, which has loose, round heads with soft leaves (think Bibb). These generally have a fairly mild flavor and tender texture.

Of course, in today's crazy mesclun world, many of the things we eat as lettuces aren't really that at all. Some are endives, which are closely related to lettuces but tend to be coarser in texture and more bitter in flavor. There are about as many endives as there are lettuces. Belgian endive (*Cichorium intybus*), sometimes called witloof chicory, is the one that is usually forced into those long, white, tapered heads. Forcing really does raise farming almost to the level of industry — or at least high craft. Belgian endive plants are started in the field. After they leaf out, they are dug up, the green tops are trimmed, and the bases are replanted in a warm, dark room. This prevents the plant from developing chlorophyll, which would turn it green and change the flavor.

Another leafy endive (*Cichorium endivia* var. *crispa*) goes by the name of salad chicory. It has curly, deeply notched leaves that are dark green on the outside, brightening to pale yellow or almost white at the center. (High-flown frisée is a slightly different variety of the same family.) This leafy endive is sometimes mistakenly called escarole. True escarole (which is another variety of *C. endivia* — *latifolium*) has broad leaves arranged in a loose head looking much like a fleshy type of leaf lettuce.

Then there is the multitude of radicchios, a noisy Italian family of *C. intybus*. The round iceberg lettuce–looking variety, Chioggia (officially, *rosa di Chioggia*) is the most familiar, but there are several others. Castelfranco forms looser heads, more like Bibb lettuce than iceberg, and is a gorgeous pale green mottled with red flecks. It looks something like an old-fashioned cabbage rose of a peculiar hue.

Treviso is a type of radicchio with a long, tapered head that is almost always used for cooking. (All three names — Chioggia, Castelfranco and Treviso — come from towns in the Veneto.) One traditional form of Treviso, called *tardivo*, needs to be hand-forced, the way Belgian endive is: plants are grown in the field, then cut and stored in the dark, where they continue to grow without developing chlorophyll. This coincidence of techniques is not accidental. The method for forcing *tardivo* was perfected by a Belgian endive grower, Francesco van den Borre, in the 1860s. The leaves of *tardivo* grow on long, white, spidery stems. It looks like some kind of exotic lily and is incomparably creamy and sweet when cooked. Because growing it takes so much work, it is rarely seen in the United States, and when you do find it, it is usually expensive.

In the 1950s plant breeders came up with another form of Treviso that doesn't require transplanting. It is called *precoce* (precocious) because it has the added advantage of maturing earlier than *tardivo*. Neither quite as sweet nor as creamy as *tardivo*, *precoce* is still very good when cooked. It is also more commonly available in the United States; look for something that resembles a big, red Belgian endive.

Lots of the greens we eat in salads today are the immature leaves of young vegetables — beets, kale, spinach, various mustards and swiss chard. These are generally tender and vary in taste according to the vegetable from which they come. Anise-flavored shiso and mustardy tatsoi and mizuna are the leaves of Asian vegetables. Salad mixes frequently contain herbs, such as chervil, flat-leaf parsley and basil. Some of the greens are what in less enlightened times would have been called weeds. Is there a gardener alive who has not cursed dandelions and purslane? Although these plants may be invasive, their peppery, slightly bitter leaves are welcome additions to mixed salads. And although it's hard to think of violet-scented mâche and spicy arugula as weeds, if you've ever planted them and let them go to seed, you know that they can make a dandelion look passive. At one point, such greens caused much hilarity among not-yet-fine diners who derided them as "lawn clippings." Fast growing and easy to harvest, they are turning into a big business. Since 1995 the acreage of "spring mix" in Monterey County alone has gone from about 500 acres to more than 12,000. You could say the lettuce industry is moving from widgets to weeds.

WHERE THEY'RE GROWN: Lettuces are grown up and down the Central Valley of California, extending into the southeastern part of the state and the area around Yuma, Arizona. They are also grown on a less industrial scale around the country, as they are one of the most popular farmers' market items.

HOW TO CHOOSE: It's no great mystery when lettuce starts to wilt. With head lettuces, make sure they are heavy for their size. In heat spells, the heads get lighter.

HOW TO STORE: Keep lettuce tightly wrapped in the refrigerator. Don't wash it until you're ready to use it. Moisture will break down the leaves faster than anything except heat. Sometimes you'll get lettuce from the market that is dripping with water from overactive misting in the produce section. If this happens, stick a dry paper towel in the bag with the lettuce to absorb any excess moisture. Refrigerate in the crisper.

HOW TO PREPARE: Tear lettuce into bite-size pieces. If you're going to serve it right away, you can cut it with a knife, but many types of crisphead lettuce will brown at the cut edges if you do this too far in advance.

ONE SIMPLE DISH: Few dishes are simpler than a tossed green salad, but that doesn't mean it is easy to get one right. Here's the easiest basic technique for a perfect salad. Cut a garlic clove in half and rub it all over the inside of a metal, glass or ceramic salad bowl. (Wooden salad bowls are attractive, but they absorb oil, which almost immediately goes rancid, flavoring every salad you make thereafter.) Place the washed and thoroughly dried leaves in the bowl (one of those salad spinners is a great cheap tool). Drizzle over just enough oil to lightly moisten the leaves when you toss them. There should not be any oil left in the bottom of the bowl. Sprinkle with good vinegar — red wine, champagne and sherry all have their attributes — and toss again. The classic proportion of oil to vinegar is 3 to 1; taste and see what works best for you. Finally, sprinkle with salt, toss once more and serve.

PRECUT VEGETABLES

You chop up some vegetables, put them in a plastic bag and stick them in the refrigerator. You come back the next day, and they're wilted. You come back the day after that, and they're halfway to rotten. And yet you go to the grocery store and see bags of precut greens and vegetables that look nearly perfect. What's the deal?

The packaging of precut vegetables is one of the most amazing advances in technology in the produce world, the result of years of research and cooperation between scientists who study plants and plastics.

Remember that vegetables continue to respire, or "breathe," after they've been harvested: they take in oxygen, and they give off carbon dioxide and ethylene. The rate at which they do this determines how quickly they spoil. Vegetables that have been cut up deteriorate even faster than whole vegetables.

At first researchers believed that if you could reduce the amount of oxygen reaching the vegetables, you could slow down the rate of respiration and extend their shelf life significantly. This worked in part — it delayed the kinds of spoiling associated with respiration. But another problem popped up. It turns out that when deprived of oxygen, vegetables begin to draw energy from their own tissues and so start on another kind of decay.

But what if you could create an atmosphere with only a little bit of oxygen and just the right amount of carbon dioxide and ethylene? By experimenting with different kinds of plastic films for the bags, scientists found a way to do this. Different types of plastic allow different flows of different gases. By combining several films, scientists were able to create plastic bags that allow a slow intake of oxygen and a rapid venting of carbon dioxide and ethylene. This extends the life of cut vegetables for days, if not weeks.

Of course, it was not quite that simple. Each vegetable has its own ideal mixture of gases and so requires its own special type of film. In addition, some vegetables require different mixes depending on the season and on harvest conditions. Last but not least, the plastic film not only has to work as a gas filter, but it also has to be able to hold the brightly colored printing that goes with commercial packaging.

The next time you pick up a bag of mixed lettuces at the grocery store, remember that that is no ordinary plastic bag.

Consommé with Shrimp, Arugula and Lemon Zest

This recipe is based on a salad I once enjoyed in Verona. It's a fussy presentation, to be sure: you arrange a little salad in the bottom of every bowl, and then you carefully pour hot consommé over the top. But when you do that, the mingled perfumes of shrimp and arugula are absolutely heady. The consommé can be used, without the refining step, as a base for soups and for making risotto. You can prepare it and cook the shrimp a day in advance, which turns this into an extremely elegant spring first course that requires almost no last-minute preparation.

6 SERVINGS

- 2 **cups water**
- 1 **teaspoon salt**
- ¼ **teaspoon red pepper flakes**
  Generous grinding of pepper
- 18 **shell-on medium shrimp (about ¾ pound)**
- ¾–1 **cup loosely packed baby arugula leaves**
  Grated zest of 1 lemon
- 10 **cups boiling Shellfish Consommé (recipe follows)**

Bring the water, salt, red pepper flakes and pepper to a boil in a small saucepan. Cook for 10 minutes. Add the shrimp, reduce the heat to low and cook until the shrimp are bright pink and heated through, about 5 minutes. Drain the shrimp, rinse with cold water and peel.

Toss together the arugula and lemon zest. Divide evenly among six wide soup or pasta bowls. Arrange 3 shrimp on top of the arugula in each bowl and take the bowls to the table.

Gently ladle the boiling consommé into each bowl, pouring from the edge rather than over the top to avoid disturbing the salad.

Shellfish Consommé

MAKES 10 CUPS

- 2 tablespoons olive oil
- 2 tomatoes, seeded and chopped
- 2 garlic cloves, thinly sliced
- 1 onion, thinly sliced
- 1 carrot, thinly sliced
- Stems from 1 bunch parsley
- 1½ pounds shell-on medium shrimp (preferably with heads)
- 1 pound mussels
- 10 cups water
- 2½ teaspoons salt
- Generous grinding of pepper
- 3 large egg whites, or more if needed

Heat the oil in a medium pot over low heat. Add the tomatoes, garlic, onion, carrot and parsley stems. Cook, covered, until the vegetables are soft, about 10 minutes. Stir from time to time to prevent sticking.

Add the shrimp and mussels and stir to coat. Add the water, salt and pepper. Bring to a simmer over medium-high heat; as soon as gentle bubbles form on the surface, reduce the heat to low. Cook for 35 minutes.

Drain the broth through a strainer, then cover and refrigerate until cool. Reserve about ¼ pound of the cooked shrimp and refrigerate. Discard the remaining solids.

To clarify the consommé, remove the broth from the refrigerator and discard the fat on the surface. Heat the broth in a large saucepan or small stockpot over medium-low heat. Meanwhile, shell the reserved cooked shrimp. Discard the shells and put the shrimp in a food processor. Add the egg whites and puree until you have a light pink foam, about 30 seconds.

Whisk the shrimp mixture into the broth and continue to heat, whisking steadily, until it boils, then reduce the heat to a simmer. Keep whisking slowly but constantly. After about 30 minutes, small bits of egg white will begin to accumulate into larger pieces about the size and shape of snowflakes. Stop whisking and let the broth continue to heat. The egg white bits will eventually collect into a moist "cap" on top of the broth. Poke a hole in the center with a spoon to allow the broth to bubble without overflowing. The broth should be in constant gentle motion, not boiling hard. Cook for 1 hour without stirring.

Using a slotted spoon or Chinese wire skimmer, very gently lift the cap from the broth and discard. Ladle the broth into a chinois if you have one. Otherwise, line a strainer or colander with moistened paper coffee filters. Pour the broth through gently, ladling the liquid against the sides and moving from spot to spot to avoid pushing any egg white through the strainer. The finished consommé should be completely clear. If it is still cloudy, repeat the clarification process, using 3 more egg whites but no shrimp. If the flavor lacks intensity, reduce the consommé over high heat for 5 to 10 minutes.

Crab Salad with Avocado and Peppery Greens

Sweet crabmeat, rich avocado and crisp peppery greens: this is one of those dishes that people just can't stop eating. If you are deprived of Dungeness crab where you live, you can substitute picked-over blue crabmeat.

10 SERVINGS

- 3 heads Belgian endive
- 2 bunches watercress
- 1½ teaspoons Dijon mustard
- 2 tablespoons champagne vinegar
- ½ teaspoon salt
- ½ cup olive oil
- 2 tablespoons snipped fresh chives
- 2 avocados
- 1 pound picked-over Dungeness or blue crabmeat (about two 2½-pound crabs, cooked and cleaned)

Trim the dried-out bases of the endive. Cut each head into quarters and cut away the solid core. Slice the endive lengthwise into ribbons as thinly as you can. Trim any tough stems and faded leaves from the watercress and tear into bite-size pieces. (The dish can be prepared to this point up to 1 day ahead and refrigerated in a tightly covered container.)

When almost ready to serve, prepare the vinaigrette by blending together the mustard, vinegar, salt, oil and chives. You can do this in a small blender jar or by shaking vigorously in a small, tightly covered jar.

Combine the endive and watercress in a large bowl and dress with about two thirds of the vinaigrette. Toss to coat lightly and thoroughly combine the endive and watercress. Arrange in a layer on a large platter.

Pit and peel the avocados and cut them into a large dice. Place the avocados in the large bowl and dress with half the remaining vinaigrette. Stir gently to coat lightly without breaking up the avocado, and spoon in a single layer over the watercress and endive.

Place the crabmeat in the large bowl, add the remaining vinaigrette and toss gently to coat lightly without breaking up the crab pieces. Arrange the crab on top of the avocado and serve.

Strawberries

· · · · · · · · · · · · ·

Several springs ago I was desperate to get my hands on some fresh wild strawberries. Unless you are a star chef with a secret supplier who hand-carries them to your back door, this is not an easy thing to do. In the first place, there aren't many farmers who grow them anymore. The ones who do tend to have only a few because the plants take so much labor and bear so little fruit. The season is vanishingly short, and the fruit is incredibly fragile. This last turned out to be the biggest sticking point.

After much research I actually did track down someone who had them, in hand and in season, but he was several hundred miles away. I told him that I would be happy to pay for shipping, but he refused. They were too delicate to ship, he said. I would pay for overnight. No way, he said, they'd never get to me in decent shape. I persisted: I wouldn't hold him responsible for any less-than-perfect berries. Finally, he caved in — probably just to get me off the phone. The next day, a big box arrived. I opened it, and there, nestled among shipping materials, was a smaller box. I opened that, and wrapped in a mound of tissue paper was a tiny pint-size box. I opened that and found the most fragrant jam I've ever smelled. Even with all that care, the berries had been smashed beyond recognition.

And therein lies the paradox of the strawberry. In its wild state, it is a highly seasonal, wildly flavorful fruit that is as fragile as a soap bubble. Yet in our passion for it, we have managed to turn this dreamy berry into a year-round staple as resilient as Styrofoam and only a little more flavorful. It wasn't so long ago that strawberries were a food you anticipated all through the winter and then gorged on in a brief frenzy that was a ritual of spring. Today it's a year-round garnish, the parsley of the breakfast plate. You can buy fresh, American-grown strawberries at least eleven months out of the year. More than 80 percent of them come from California, which produces more than a billion pounds in total. That means the strawberries have to be able to withstand a four-day truck ride to make it to the East Coast.

To provide a year-round supply, farmers harvest strawberries from one end of the state to the other, beginning in San Diego and Orange County in the south right around Christmas and gradually moving north as the season progresses and the weather warms, finishing up around Watsonville, just south of San Francisco, around Thanksgiving. In a good year, one with a mild and extended summer, strawberries never go out of season.

And yet finding a berry with true flavor — the kind that stops you in your tracks when you taste it — just keeps getting harder. There is a solution, though. Despite the fact that California has an overwhelming commercial edge, strawberries are one of the most widely grown farmers' market fruits. And this is one case where the old "buy local; buy seasonal" mantra really pays off.

Locally grown berries, which don't have to make a cross-country trek before you can eat them, will almost always be juicier and more flavorful than their commercial counterparts — even if they're grown from the same variety. And fortunately, strawberries are almost uniquely fitted for small farmers. Although they demand a lot of extremely tedious handwork to grow, they offer among the highest cash returns to farmers.

So lucrative are strawberries that even in these days of consolidation and ever bigger farms, it's possible for a grower to make a living on less than ten acres. That's why strawberries are the overwhelming favorite of urban farmers — those hardy souls who practice agriculture in the small, often temporary open spaces found in cities. You can find farmers growing strawberries on a couple of acres under power lines, and you can find them tending their fields on land that is being cleared for housing developments (in these cases, strawberry fields are definitely *not* forever).

This friendliness to small-scale, transient farming is the reason behind one of the more interesting chapters in the history of American strawberry farming. At the turn of the century, when the California strawberry industry was just becoming established, it was heavily populated by Japanese immigrants. The labor-intensive, highly profitable farming was ideal for growers with extended families. Furthermore, these growers were able to turn another of the strawberry's weaknesses to their advantage. Strawberries are susceptible to all kinds of pests, many of which were not controlled until after the advent of chemical pesticides after World War II. Verticillium wilt is particularly vexing. Until the 1950s the soilbound fungus that causes the wilt would kill any strawberry field that remained planted in the same location for more than a couple of years. This vulnerability forced strawberry growers to be a highly mobile lot, and most of them rented land rather than owning it.

The situation was ideal for Japanese American growers, because in the early part of the twentieth century, it was illegal for them to own land in California. These growers turned two negatives into a positive by focusing on strawberries. A survey taken in 1910 found that almost 80 percent of the strawberry growers in Los Angeles County were Japanese American. When the Central California Berry Growing Association, the first strawberry marketing co-op, was founded in 1917, the bylaws required that half of the board of directors be Japanese American — an extraordinary move during a period so virulently anti-Japanese.

Certainly, today's small strawberry growers do not face anywhere near the same hurdles as the Japanese American farmers did a century ago. But that is not to say that their lot is a walk in the park. In particular, they have to deal with sometimes cranky neighbors, for whom the realities of agriculture — dust, early mornings, lots of workers coming and going, occasional spraying — do not quite mesh with their idea of the good life. But because strawberries are so valued by fruit lovers — especially good strawberries, picked ripe and shipped only across town rather than across the country — these farmers are able to earn enough to make it worthwhile.

When you do get those perfect berries, remember that they almost always taste best uncooked. The red color of berries comes from the pigment anthocyanin, which is not heat stable. If you cook strawberries by themselves, that lovely crimson color will turn to a bruised purple. But acidity will stabilize the pigment, so add some lemon or orange juice (or bake them with rhubarb), and the color will remain red. You can "cook" strawberries without heat, though. Sugar draws moisture out of strawberries and mixes with the extracted juice to form a delicious sauce. In some cases, this can be bad — if you want the berries to remain slightly firm, don't sugar them too far in advance of serving, or they'll go limp. In other cases, the sugaring is a big help — sugar strawberries for ice cream well in advance of freezing, and because of the extracted moisture, you won't end up with ice cubes in your ice cream.

WHERE THEY'RE GROWN: The vast majority of commercial strawberries are grown in California. But strawberries are one of the leading "small-farm" crops around the country. Varieties that are grown for the local market — without the necessity of shipping — are almost guaranteed to be better than most commercial berries.

HOW TO CHOOSE: There are a lot of little indicators of strawberry quality, but the most important is probably the simplest: smell. Great strawberries have a distinctive candied aroma that you can't miss. Beyond that, the berries should be completely red (the exact shade of red will depend on the variety); avoid any with white tips. The green hull should look fresh, not dried out. The berries should be glossy, without any matte spots where the flesh has started to break down. Always look at the underside of the berry basket — that's where crushed berries may be hiding and where spoilage will start. It's not at all uncommon to pick up a basket of berries that are beautiful on top but are as gray and fuzzy as a freshman dorm refrigerator underneath.

HOW TO STORE: This is a tough one, because refrigerating damages the flavor of strawberries, but the fruit is so tender that not chilling will lead to rapid spoilage. The best solution is to buy berries from a local farmer and eat them the same day without putting them in the refrigerator. Failing that, transfer the berries to a plastic bag (to prevent excessive drying) lined with a paper towel (to absorb excessive moisture) and refrigerate them.

HOW TO PREPARE: Don't rinse strawberries until just before you're ready to use them; the moisture will speed decay. And don't remove the green hulls until after you've rinsed the berries. Those caps prevent the berries from soaking up too much water. Once they've been rinsed, gently blot them dry with a paper towel.

ONE SIMPLE DISH: Whisk together a bottle of light red wine or rosé and a cup of sugar. Add a split vanilla bean. Cut up 2 pints of strawberries and add them to the wine mixture. Refrigerate for at least 2 hours or overnight. Ladle the strawberry soup into bowls and serve each with a scoop of vanilla ice cream and a crisp cookie.

Ole's Swedish Hotcakes
with Quick Strawberry Compote

Of all the breakfasts in the world, this recipe, adapted from one pre-pared at the Little River Inn just south of the town of Mendocino, California, is one of my favorites. The pancakes are served with a big spoonful of strawberry compote in the center. To really gild the lily, you can top that with a spoonful of whipped cream.

4 SERVINGS

12 tablespoons (1½ sticks) butter
 1 cup all-purpose flour
 1 teaspoon baking powder
 1 teaspoon sugar
 ¼ teaspoon salt
1½ cups milk
 ½ cup half-and-half
   Grated zest of 1 orange
 3 large eggs, separated
   Quick Strawberry Compote (recipe follows)

Melt the butter and let it cool slightly.

Meanwhile, stir together the flour, baking powder, sugar and salt in a large bowl. Whisk in the milk, half-and-half and orange zest. The mixture will be very liquid; don't worry.

Whisk in the egg yolks. This will thicken the batter slightly. In a separate bowl, beat the egg whites until soft peaks form and stir them gently into the batter. (You don't need to fold them; the batter is not that delicate.) This will thicken the batter to about the consistency of a good home-made eggnog. Whisk in the melted butter. (The recipe can be made ahead to this point and refrigerated, tightly covered, overnight.)

Heat a nonstick skillet over medium-high heat until drops of water skitter across the surface. Slowly pour ½ cup of the batter in the center of the skillet, forming as much of a circle as you can. (Using a ladle or measuring cup with a lip makes this easier.)

Cook until the bottom of the pancake is lightly browned and the top begins to look slightly dry, about 3 minutes. Flip the pancake and cook until it feels somewhat firm when pressed lightly in the center, about 2 minutes more.

Remove from the pan and keep warm in a 200-degree oven as you continue with the rest of the batter. Serve 2 pancakes per person, with a generous portion of compote.

Quick Strawberry Compote

MAKES ABOUT 1 CUP

½ **pound strawberries, rinsed and hulled**
¼ **cup sugar**
1 **tablespoon fresh orange juice**

Place the strawberries, sugar and orange juice in a food processor and pulse 4 or 5 times just to chop the berries small. Do not puree.

Transfer the mixture to a small nonstick skillet and cook over medium-high heat until it begins to thicken, about 5 minutes. Set aside until ready to use. Serve warm or at room temperature.

Strawberry Preserves

By preparing preserves in small batches, the jam will cook quickly enough that the fruit retains its fresh taste. This recipe works best by weight. (How else would you know if you were a few strawberries short of a pint?) Use equal amounts of fruit and sugar. We've listed approximate volume measures if you don't have a scale (2 pints of strawberries weigh about 2 pounds).

If you haven't made jam before, you'll want to familiarize yourself with the basics on pages 109–11. You do need to sugar the berries the night before.

MAKES FIVE 8-OUNCE JARS

- 2 pounds strawberries, rinsed, hulled and cut into bite-size pieces (about 8 cups)
- 2 pounds sugar (about 4 cups)
  Juice of 1 lemon or orange

Combine the strawberries and sugar in a large pot and heat slowly until the juices are clear, about 5 minutes. Remove from the heat and stir in the lemon (or orange) juice, then cover loosely and let stand overnight.

The next day, get everything ready for canning. Bring a large pot of water to a boil and sterilize 5 sets of jars and lids, about 5 minutes. Turn off the heat, but leave the jars and lids in the hot water until you're ready to use them.

Heat 2 cups of the strawberries and juice in a large nonstick skillet over medium-high heat. When the strawberries start to simmer, cook, stirring often, until the preserves test done (see page 111), 3 to 5 minutes.

Ladle the jam into the sterilized jars, filling to within ¼ inch of the rims. Cover each jar with a lid and fasten the ring tight. Set aside and repeat with the remaining strawberries and juice.

Seal according to instructions on page 111.

PRESERVES

"Nobody makes jam at home anymore, except for shut-ins and little old ladies," a long-ago editor once told me. I was thinking about that Sunday as I whipped up a mess of strawberry preserves after breakfast, in between going to the farmers' market and getting ready to watch the Lakers. Granted, home preserving has an image about as hip and sexy as a gingham apron. But hip and sexy is a passing pleasure, especially when compared with the flavor of a spoonful of my strawberry preserves smeared on a piece of hot buttered toast. And don't even get me started on my Elephant Heart plum or nectarine and rose geranium jam.

It is surprising that in this do-it-yourself world of cooking, where people brag about making their own bread, fresh pasta and chicken broth, jam making is still so little regarded. That's especially true considering this technique, which makes creating your own jams and preserves about as complicated as cooking a quick pasta for dinner. All you need is a big pot, a nonstick skillet, jars and a scale. Oh, and fruit. But just a little bit.

The problem with most preserving is that it requires such large quantities. This requirement does have the advantage of making enough preserves to take you through a hard midwestern winter. However, that benefit is outweighed by the finger cramps you get when paring a bushel of fruit, the amount of time required to cook it all and the sheer uncertainty involved when preparing jam in such large batches. (Is it done now? Now? Now?)

With this abbreviated method, a couple of pints of strawberries turn into four or five jars of jam in less than 20 minutes of cooking time. You have to let them sit overnight — to let the sugar pull all that sweet juice out of the fruit — but how much trouble is that? I first tried this method a couple of years ago at the suggestion of a friend, cookbook author Sylvia Thompson. Before long, just about every time I went to the market, I'd come back with a couple of pounds of something to turn into jam. This basic technique will work with everything from strawberries to peaches and plums.

To start, wash the fruit and cut it into bite-size pieces. Weigh it, then put it in a pot with an equal weight of sugar.

Bring the mixture to a boil and stir until the juices are clear. Then set it aside for several hours or overnight. Ladle a couple of cups of fruit into a skillet and bring it to a boil. Cook until the mixture has set, 5 minutes or less, and you're done.

When you're working with only 2 cups of fruit, you can tell by the feel when the mixture has set. Suddenly, it's thicker and smoother as you're stirring. As a longtime preserver, I am still startled that it cooks so quickly. In fact, the first batch usually winds up slightly overcooked. ("Let's go a little longer; it can't be done already!" But it is.)

You'll notice that this jam is a little softer than what you may be used to. That's because it's made without artificial pectin, which results in a very firm set but which also requires more sugar than I like to use.

In fact, according to the Food and Drug Administration (FDA), you can't really call this jam. FDA regulations on product identity require jams, jellies and preserves to contain at least 65 percent sugar. This jam has only 50 percent sugar and can be made with even less. That leaves it with a much fresher taste and a brighter color.

You can easily store the four or five jars of jam this recipe makes in the refrigerator. They'll last a couple of weeks, and then you'll be ready to make some more. But if you want to can the jam so that you can store it without refrigeration, you'll need to devote another 15 minutes to the process.

Despite the mystique surrounding it, canning isn't hard. The main thing is making sure everything you use is sterilized — easy enough to do when one of the requirements is a big pot of boiling water. Boil the jars and lids for 5 minutes, and you're ready to go. Ladle the jam mixture into the jars (a wide-mouth canning funnel isn't necessary but will certainly make things easier). Put the lids on the jars and tighten down the bands. Now put the jars in the boiling water, making sure they're completely submerged. (If you have a pasta insert — the deep basket that comes with a pasta pot — you can add the jars and remove them much more easily.) Cook for 10 minutes and then lift the jars from the bath. After a couple of hours, check to make sure you have good seals. (The merry pinging you should hear during the cooldown will be the lids popping from the vacuum forming.) Press down firmly on the center of each lid; there shouldn't be any flex. If the lid flexes, repeat the canning process.

Testing Preserves for Doneness

If you've had trouble in the past telling when jam has jelled, try this method. To make certain the jam is cooked, dip a rubber or plastic spatula into the mixture, then lift it out. When the mixture no longer flows from the side at one point but instead begins to come off at multiple points or in sheets, it's done. Or use the plate test: Chill a plate in the freezer for 10 minutes. When you think the jam is ready, spoon a drop of liquid onto the plate. If it doesn't run when you tilt the plate, you're done.

Strawberries and Oranges in Basil Syrup

Strawberries have an undeniable affinity for the cinnamon-spicy flavor of basil. This recipe can be prepared several hours in advance, but if you store it too long, the berries will soften and lose their texture. Serve this with sugar cookies to dunk in the leftover syrup.

8 SERVINGS

- 1½ cups water
- ¾ cup sugar
- ¼ cup thinly sliced fresh basil leaves
- 2 pounds strawberries
- 4 navel oranges

Prepare the syrup by bringing the water and sugar to a boil and then cooking just until clear, 2 to 3 minutes. Let the syrup cool very briefly, then add the sliced basil. Set aside to steep and cool while you prepare the fruit.

Rinse the strawberries, remove their caps and cut them into quarters if normal size or sixths if very large. Peel the oranges, being careful to remove all of the bitter white pith. Working over a bowl to catch any juice, cut the oranges into sections, leaving behind all the tough skin that separates the sections.

Add the strawberries to the orange sections and juice and pour the syrup over the fruit. Refrigerate for at least 30 minutes before serving.

Big Farmers, Small Farmers

THE COST OF COMPROMISE

The world of agriculture is defined both practically and aesthetically by two poles: those farmers whose aim is to grow the greatest possible amount of food at the lowest possible price, and those whose goal is to grow the greatest quality of food no matter what it costs. Both are necessary. It would be a poor state of affairs if our grocery stores were filled with nothing but $6-a-pound peaches, no matter how exquisite they might be. And, of course, the opposite is equally true: what good does it do you to have all the peaches you can eat if none of them tastes like anything? Most growers fall somewhere in between, but at their extremes these two schools of thought operate in almost separate universes — despite the fact that the growers themselves might be farming right next door to each other.

Nowhere is this more evident than in the stone fruit belt of California's San Joaquin Valley during the middle of summer. Almost all of the nectarines and about 60 percent of the peaches grown in the United States come from California. And about 90 percent of those are grown between Fresno and Visalia, a stone fruit belt no more than fifty miles wide encompassing the southern half of Fresno County and the northern half of Tulare County. That's a hefty crop. Peaches alone account for almost $250 million a year, and nectarines are worth another $200 million. But this is an area almost uniquely suited to the growing of peaches and nectarines. In the winter, the heavy cold air of the Sierra Madre range slides downhill to rest along the banks of the Kings River. This phenomenon produces the area's dense tulle fog, infamous among drivers, but it also gives the trees the chill they need to go dormant and get their off-season rest. In the summer the hot, dry weather is perfect for ripening fruit with a minimum risk of spoilage. Small towns — once market centers — are scattered every seven or eight miles along the railroad tracks that run up and down the valley's spine.

Back in the 1890s, when agriculture was just starting in the area, that was about as far as a farmer could carry his crop, sell it and still get home in a day. Today the railroads have been replaced by almost impossibly straight highways. Scattered along the side of the road, among the vineyards and orchards, are 1960s brick ranch-style ramblers, along with the occasional Nouvelle Colonial stucco mansion. Here and there you can spot old plains-style farmhouses, some of them quite grand, given away by the telltale "tank house" out back that was originally used for storing water for household use.

Fruit packing sheds dot the area, and the quickest way to kill any romantic notions you might have about the nature of big-time agriculture is to visit one in the middle of the summer harvest. The ungodly din, the frenzied activity, the seemingly endless supply of fruit — all your imagined bucolic scenes of Farmer Brown in his overalls handpicking his peaches will be swept away in an instant. This is where the industrial side of farming becomes clear — food grown on a massive scale and intended to be sold at a bargain price. Here is the source of almost all of the fruits and vegetables in your grocery store, and it is impossible to understand the modern produce section without understanding the differences between it and your peachy nostalgic dreams. These packing sheds tend to look much like any other light manufacturing plant: about the size of a football field, roughly two stories tall, made out of prefabricated metal and prefabricated concrete. From the outside, it can be hard to tell whether they're for peaches or water pumps. Except for the lingering perfume of fruit, there are few clues inside either.

The machinery looks like some gigantic Rube Goldberg invention, a mechanical contraption so incredibly complex that it seems it couldn't possibly serve any discernible purpose. But then it swings into action. At one end is a massive garage door, which opens to allow the entry of a truck loaded with field bins, each about the size of a large Dumpster. A lever locks onto a bin and gradually, with a loud mechanical groan, lifts and tilts it, spilling its load. Onto the padded conveyor belt tumble more nectarines than you have probably ever seen in one place, each of them perfectly round and gleaming candy-apple red. One piece of fruit after another, each seemingly more perfect than the last, shines and catches the light as it rumbles on its cushioned way from field bin to shipping box. The fruit couldn't be more uniform if it were manufactured.

And in a sense, it was. This fruit was designed — you can almost use that word literally — to be grown, packed and sold on a massive scale. The trees are bred to be disease-free and heavy bearing. They are also selected to be short and compact, which permits easier picking (and reduces legal issues — tall trees require ladders, which increase an employer's liability). They are heavily fertilized and watered, which creates the largest volume of the biggest, glossiest fruit.

Harvest is a combination of military operation and invasion by locusts. Starting at one end, pickers — almost uniformly Hispanic males — move from tree to tree, nearly sprinting, silent as ghosts. Even though the morning is already warming, they are dressed in sweats to protect them from the trees and the sun, and they wear baseball caps with short, dangling capes in back like the French Foreign Legion. There is no playing around; their concentration on the job is almost complete. They will eventually pick every piece of fruit they see, tossing nectarines over their shoulders into the big canvas buckets they carry like golf bags. When the buckets are full, they are emptied into the field bins, and the fruit is trucked to the packing sheds.

All of the fruit on a tree doesn't come ripe at exactly the same time, of course, but picking is expensive, so most orchards like this will be harvested only two or three times, several days apart. (Most peach and nectarine varieties come ripe only for a period of ten to fourteen days.) A large-scale commercial orchard may cover 120 to 150 acres and include dozens of different varieties of peaches and nectarines. In one summer, it may produce more than 2,000 tons of fruit. Although there is considerable variability depending on the climate and the year, a little more than 10 tons per acre would be considered average for peaches, about 7.5 tons per acre for nectarines.

The fruit may be highly colored, but it is not yet ripe. The red coloring is a genetic trait that shows at a very low level of maturity. A ripe nectarine put through the packing process would turn to jam. Instead, the fruit is picked at a stage of physical maturity at which most of it can, if handled correctly, develop pretty good flavor.

Fruit is graded and sorted according to background color and softness of flesh. The lowest grade is "utility" and is reserved for fruit that is edible but may be cosmetically damaged. The next is "U.S. mature." Then there is "California well-matured," which ostensibly is harvested at a higher order of maturity. These days, about 90 percent of the peaches and nectarines harvested in the state qualify for this standard, so you be the judge as to how meaningful it might be. Fruit can also be labeled "tree-ripe," but this is essentially meaningless, as the maturity standards are the same as for California well-matured.

At the packing shed, the first thing that happens is the fruit is sorted. The bins are emptied, and the nectarines roll down the conveyor belt in what seems like an endless river. These nectarines don't necessarily come only from the orchard that was just being picked, but may come from several others as well — farms that are too small to have a packing shed of their own or that simply don't want the bother. Here all the fruit is combined into one lot, sorted only by variety.

Women workers line both sides of the belt, culling any fruit that is obviously damaged. Rejected nectarines will be used for animal feed or fertilizer. The fruit that is left runs under an electric eye that sorts it by size. As each nectarine reaches the gate for its size, it is shunted through into the hands of one of the waiting women, who carefully places it in a packing case. Peaches and nectarines are typically packed into a two-layer cardboard tray container to a total weight of roughly twenty-two pounds. These containers are then stacked on wooden pallets and toted by forklift to be cooled.

This cooling stage is critical. At harvest the internal temperature of nectarines can be in the 90s. If left unattended, they would spoil in less than a day. The ideal storage temperature for peaches and nectarines is *right at freezing* — 32 degrees (because they contain so much sugar, peaches don't freeze until they get between 27 and 30 degrees). Much warmer than that, and the fruit enters the perilous "chilling injury" zone — roughly 35 to 45 degrees. The reasons behind the harmful effects of cold storage within this temperature range have not been fully explained, but the symptoms are well known: a cottony, dry texture and an absence of flavor. As little as a day spent at these temperatures can be enough to ruin a piece of fruit.

Different sheds cool their fruit in different ways. Probably the most common method is forced-air cooling, basically putting the fruit in a refrigerator the size of a handball court. The next most popular is hydrocooling, streaming ice-cold water over the bins, like putting the fruit under a giant cold shower. The Rolls-Royce of chilling techniques is vacuum cooling, but this is so expensive that it is used only for extremely perishable items such as herbs and leaf lettuces. The fruit is placed in a sealed container, and all of the air is pumped out. When the atmospheric pressure gets low enough, some of the moisture from the fruit begins to evaporate and cool the fruit.

Some shippers have begun using a new cooling process called "preconditioning," which shows promise in eliminating chill injury. Developed by the University of California at Davis, it calls for the fruit to be cooled only to between 68 and 77 degrees and then to be held for twelve to thirty-six hours to allow the fruit to begin the ripening process. The fruit is then chilled to below 34 degrees for storing and shipping. The fruit is held in giant refrigerated rooms, with pallets stacked fifteen to eighteen feet high, until it is ready to be shipped.

With the fruit carefully tucked away for its nap, the action in the packinghouse shifts to the glassed-in offices that encircle the upper level. This is where the salespeople work the phones, frantically scrambling to sell all of the fruit that has just been harvested. Peaches wait for no one, even when picked so underripe, and they certainly won't wait for a better price. For that reason, in produce country it is almost always a buyer's market, and everyone knows that. Conversations are rarely about "How sweet are those peaches?" They're almost always along the lines of "Is that the cheapest you can sell them?" On the other end of the phone are a wide variety of buyers. Some represent large grocery store chains; what they take will be trucked to their warehouses and then distributed to individual markets. Some are wholesalers, who will bring the fruit to large collective warehouses called "terminal markets," where individual stores can buy them in bulk. Some are specialty companies that buy the fruit and then ripen it further before selling it to high-end users (either markets or restaurants).

At each step of the way, someone is making a profit. And by the time you track backward to the source of the fruit — the farmer — the prices can be shockingly low. Only about 20 cents of every dollar you spend on peaches at the grocery store winds up in the pockets of the people who grew them. (To put this in perspective, according to one university study, it costs about $11,000 an acre to grow, pick and pack peaches in the San Joaquin Valley, including amortizing the cost of the land and equipment. At an average yield of ten tons per acre, a grower whose peaches are selling for $2 a pound at retail will get about 40 cents of that, meaning that he will lose $3,000 per acre.)

Growing better-tasting fruit involves both money and risk. For a farmer every harvest is a race against time, weather and misfortune. Every day the fruit hangs on the tree is another day it might rain or blow, another day for bugs or birds or some other calamity to find it. Entire orchards of fruit have been lost because a grower gambled on waiting one more day and got hit by a hailstorm, wiping out a whole year's work. The goal of a commercial farmer is to harvest the fruit as soon as it reaches a minimum quality standard. This is the way commercial agriculture has worked since at least the end of World War II. These farmers don't see themselves as talented artists creating sensual masterpieces. They are technicians, and they are proud of how well they are able to feed so many people so cheaply. Given the realities they face, it really is a minor miracle that our fruit is as good as it is.

In the world of commercial farming, this kind of packinghouse stands at the pinnacle. It is a grower/packer/shipper, an all-in-one operation. It grows the fruit, packs it and has a dedicated sales crew to sell it. There are also individual growers, who farm the fruit and then contract to have it packed and sold. And there are packinghouses that do no farming; they pack and sell fruit from individual growers. Regardless of category, it is extremely rare — nearly impossible — for fruit grown by a single farmer to reach the supermarket without being mixed with someone else's. As far as we are concerned, the farmers are faceless and nameless. So what is the incentive for someone who grows terrific peaches to invest in the extra work and money that it requires to do this, when his fruit is going to end up being combined with that of his neighbor, who might not necessarily have the same standards? There really doesn't seem to be one. This is one of the great dilemmas of commercial agriculture: there are significant rewards for growing more fruit (in fact, it is almost required), but there are precious few for growing better fruit.

For a gifted farmer to reap the benefits of his talents and efforts, he is almost forced to go outside the normal supply chain. The most common escape route is the farmers' market, which provides a grower with both blessings and curses in roughly equal proportions. If you want character — both in fruit and in farmers — these markets are certainly the place to go.

Two of the biggest stars at the Santa Monica farmers' market are the stone fruit growers Art Lange and Fitz Kelly, small farmers whose orchards are practically next door to each other just south of Fresno. Bite into one of Lange's Snow Queen white nectarines, and the flavor is enough to make you gasp. The first impression is of powerful syrupy sweetness. Then comes a tart tang that gives the sugar some backbone. Overriding everything is a mix of complex flavors, both floral and fruity, so mouth-filling it seems almost meaty. The fruit is so ripe that the juices drip down your chin; so ripe that a peach practically peels itself. Much the same can be said for Kelly's Lady in Red peaches.

These days, when we consider ourselves lucky to get fruit that is simply sweet, it is easy to forget that something as basic as a peach or a nectarine can actually have the power to shock. Producing fruit like that is no accident of nature. It takes a gifted farmer, a lot of hard work and a refusal to compromise. Kelly and Lange have been growing such amazing fruit for so long that they have come to embody great farming. They remind us that growing food can be every bit the work of art that cooking it can be. Other growers have customers; Kelly and Lange have apostles. And here's the thing: Kelly and Lange farm right in the middle of the stone fruit belt. The big boys are their neighbors, farming the same type of land, often with many of the same types of trees.

Kelly is a loquacious, good-looking Irishman with an impressive head of wavy silver hair and a bluff, boyo charm. Picture a younger, healthier Ted Kennedy with a farmer's tan, perpetually clad in khaki shorts and a faded work shirt. As he bangs through the twenty-acre main orchard he's owned for more than thirty years in a beat-up four-wheel-drive convertible, he can't stop talking about the things that please him about the land, whether it's the lineage of an odd fruit tree or the red-tailed hawks and great horned owls that live in the eucalyptus island at the center of his property.

Kelly came to farming almost by accident. He was working as a carpenter, looking for a farmhouse to fix up and sell, when he first found the property. But, he says, once he stepped on the land, he couldn't leave. He bought the farm in 1972 and since then has added another fifteen acres. He lives in a two-story house he built over his packing shed. "A very smart man told me the most expensive part of any structure is the roof, so you ought to get as many floors under it as you can," Kelly says. Practicality obviously wasn't the only reason, though. Sitting on one of his decks, perched high above the surrounding countryside, you have a bird's-eye view of hundreds of acres of nearby orchards, most of them big commercial operations he calls factory farms.

Kelly stops the car to snag a low-hanging white peach off a tree limb. It's so sweet it almost tastes like a sugar cube. "Wow, we've got to test that one," he says and slams back to his packing shed to pick up his refractometer — a device that measures sugar content. It's the same tool winemakers use to tell when grapes are ripe enough to make great wine. This particular peach maxes the meter at 23 percent. (California well-matured fruit averages 11 to 12 percent. Anything over 18 percent, peach marketer Jon Rowley says, "almost goes beyond the human threshold for pleasure.") Kelly looks pleased and tells about a peach he once tested that measured 30 percent.

By comparison with their neighbors, Kelly's and Lange's orchards look downright scruffy. The trees seem to be smaller and the weeds taller. That's fine with them. Big healthy trees don't necessarily produce the best fruit, they say. Sounding like high-end winegrowers, they say that they want to stress their trees to concentrate the flavor in the fruit. Lange points out the lush green foliage of his neighbor's commercially farmed trees. "That's really beautiful," he says, "but you can only get that by using a lot of nitrogen, and that makes his fruit taste sour."

Indeed, there's little that can be more stressful than trying to grow fruit in the fine sand that makes up most of their farms. The soil is so nutrient-poor that Kelly jokes that he's almost farming hydroponically. That is one reason — in addition to sheer contrariness — that neither Kelly nor Lange is certified organic (although both use only minimal amounts of chemicals and only when absolutely necessary). Stressed to the edge of survival, these trees need all the help they can get. From time to time and in carefully measured doses, fertilizers are fed in minute quantities. Watering is treated almost as an art form, with water being applied abstemiously following a carefully worked-out, highly regimented routine. (Despite being friends and farming practically next door to each other for decades, only recently did they discover that their "secret" watering techniques are practically identical.) Most of the time, they rely on beneficial insects rather than insecticides: the bad bugs are eaten by better bugs. And those aren't weeds between the trees, but a carefully chosen blend of vetch, peas, barley, wheat, rye and wild oats that adds nutrients to the soil.

Lange, a tall man who is going a little stooped now that he's in his eighties, bought his seventeen-acre farm in the early 1970s, when he was at the University of California's nearby Kearney Agricultural Center. A weed scientist by training, Lange is one of the few farmers at any market with a doctorate in plant physiology. This tends to lend his conversations a professorial air.

While Kelly talks birds out of the trees, Lange is a gentleman of a few carefully reasoned, well-chosen words. Like Kelly, he lives in the midst of his farm. But while the Irishman's aerie is an architectural flight of fancy, Lange's couldn't be more down-to-earth. He inhabits a one-room A-frame decorated in Early Bachelor Farmer. The walls are lined with bookcases full of agricultural textbooks, and the floors are stacked with technical publications. The A-frame looks more like a disorderly cross between a campus office and a storage shed than a home. "Look at our houses, and that sort of describes both of us," Lange says.

Kelly and Lange pick their fruit nearly dead ripe, when it has already begun to soften. Lange's goes straight from the tree into a packing tray lined with a single layer of individual protective cups. When that is filled, it is taken to a truck, where another worker sorts the fruit according to size. That is the last time it is touched until it gets to market. Picking fruit this ripe entails risks even beyond those associated with packing and handling.

When you follow the picking crew, the price of this gamble becomes obvious. Workers who harvest Lange's famous Snow Queen white nectarines — which can sell for as much as $6 a pound — seem to leave fully half of the fruit on the trees. Maybe a nectarine is too small or is split (something the variety is prone to do); maybe it's been gnawed by a pest. When the fruit that does pass muster gets to the truck for sorting, what seems like another half is discarded. The closer inspection turned up a bruise, excessive russeting from the sun or a spot on the neck where it rubbed against a twig. Few of these faults would have been obvious if the fruit had been picked a week earlier.

The cost of perfection is enormous. Whereas the average stone fruit farmer in California harvests about ten tons per acre, Lange and Kelly pick only two or three. This difference in sales volume could never be recouped through normal commercial channels; it is only by direct marketing that growers can get a premium for a great product. Peaches and nectarines at many supermarkets can go for less than $1 a pound, and even good farmers' market fruit might sell for $2 a pound, but stone fruit grown by these two men fetches far higher prices. And people stand in line to buy it. Even at those elevated prices, however, the economics are tough. Multiply an average of $4 a pound by two tons per acre, and you're still barely in the black — especially when the profit is spread over so few acres.

But even setting economics aside, life as a farmers' market grower is not all artistic rigor and rustic bliss. There is a dirty, even a dangerous, practical side to it. Somebody has to get that fruit to the market, and in most cases it's the farmer. Compounding the issue is the fact that most farmers' markets start early in the morning. So, in addition to growing great peaches and nectarines, these guys have to be willing to jump in a pickup truck at the end of a long day and drive the four and a half to five hours to the big markets in San Francisco and Los Angeles. Every year in California, three or four farmers' market growers are killed in automobile accidents while making this long, exhausting trek.

For Lange every week is a test of whether he still has the stamina to stay awake. His continued presence at farmers' markets is strictly a week-by-week race against encroaching age. Highway time is not the only hurdle growers encounter in this phase of their operations. Because the early market starts usually dictate an overnight stay, farmers sometimes are forced to get creative in arranging accommodations. Some have friends in their market towns that lend them a couch, or even a room. For Lange and Kelly, the choices are often less hospitable. At dinner one night, the two spent half an hour comparing the relative merits of various residential construction sites in Santa Monica and Brentwood. Frequently, they end up spending the night in their trucks, parked in the driveway of a house that is unoccupied during remodeling.

It's a challenging business, but what keeps Lange, Kelly and other farmers of their ilk going is the notion that what they're doing is more than just a business. Sure, they have to earn enough money to keep going the next year — there are no arts grants for fruit growers — but these guys passionately believe that they're saving the very idea of great-flavored fruits and vegetables from the blanding effects of modern farming. Like the Blues Brothers, they're on a mission from a different god.

Ask Kelly about a commercial peach, and he goes practically apoplectic. "You know, I'll tell you the truth," he says. "The tomato has always been the example of what people hate about modern farming. They remember it tasting so great, and it doesn't taste like anything anymore. I honestly think the peach is going to be in that league, too. All of these factory farmers, they've got an awful lot of facts. They can tell you how many hours of sunlight a peach needs, and they do everything by the rules. But their fruit doesn't have any flavor."

By now he's nearly sputtering: "The question I always want to ask them is, 'Would you eat that, Mr. Farmer?' If the answer is no, then why do they think Harry Housewife would? Why would you want to pay for something that doesn't taste like anything?"

WHEN IT'S OKAY
TO BUY UNRIPE FRUIT

One of the biggest mistakes people make when they're shopping for fruit is assuming that what they see is what they're going to get. Many fruits will improve if you can just leave them alone for a day or two.

There is a difference between maturity — basically, the development of sugar in a fruit — and ripening — the many physical changes that involve the softening of the flesh and the development of aroma and complex flavor. In most fruits, these two processes run concurrently and stop at harvest. But with some fruits, the ripening process can continue after the fruit has been picked, provided that it has attained a sufficient level of maturity.

Generally, these are known as climacteric fruits (and this is one instance where it is important to remember that avocados and tomatoes are not truly vegetables). To the agricultural scientist, the term "climacteric" refers to the physiological point at which fruits begin to ripen. (It also refers to humans of menopausal age: those of us who are fully mature but perhaps not yet fully ripe? And certainly not yet senescent!) Climacteric fruits are those that will ripen on their own, off the tree. They won't get any sweeter; since the development of sugar is an effect of maturity, but their flesh will soften, and they will become more fragrant and complex in flavor.

Climacteric fruits are a real boon to farmers, who can harvest them when they are physiologically mature but before they have begun to soften.

Different climacteric fruits will ripen at different rates. Apples are among the most rapid. As a result, the apple industry has developed a whole host of procedures to delay the process. Slightly slower are apricots, avocados, muskmelons, plums, peaches and nectarines. Bananas and tomatoes ripen even more slowly.

There are two keys to ripening climacteric fruits after harvest. The first is temperature, which should be warm enough to encourage respiration, but not so hot as to promote spoilage. The presence of ethylene gas is also a boon. Ethylene is a natural gas given off by fruit as it ripens. In nature it serves as a kind of wake-up call: the first fruit to ripen begins to give it off, which lets the other fruit know it's time to get busy.

You can buy all sorts of gadgets to help you ripen fruits, such as plastic globes, but all you really need is a paper bag and a spot on your counter. In many cases, you don't even need the bag. It serves only to trap the ethylene gas and encourage even, rapid ripening. A bag helps most with slower-ripening fruits. For those that ripen more quickly, placing them on a plate or in a bowl will work just fine if you make sure to turn the fruit once a day or so to prevent any spoil spots from developing.

Fruits That Will Ripen After Picking

APPLES

APRICOTS

AVOCADOS

BANANAS

CANTALOUPES aka muskmelons
(but not honeydew or watermelons)

FIGS

GUAVAS

MANGOES

NECTARINES

PEACHES

PEARS

PERSIMMONS

PLUMS

QUINCES

TOMATOES

Summer

Corn

· · · · · ·

In corn as in life, be careful what you wish for. Just try finding an ear that tastes the way it used to, and you'll see what good intentions can do. For generations, Americans have worshipped a sweet corn as one of our national culinary treasures. But it was a gem with a flaw. Cooked immediately after picking, corn was superb; cooked a day later, it was much less so. Much of its flavor comes from sweetness, and that sugar converts to starch very quickly — an ear of regular corn loses half of its sweetness within twenty-four hours. So plant breeders worked to overcome that difficulty, developing new varieties, some that are much sweeter than the traditional ones, some that go starchy more slowly and some that do both.

Today these new and improved varieties are almost the only kinds you can find. Although plant breeders have inarguably succeeded in making corn sweeter, it's not altogether clear that we're better off for it. There's a lot more to corn flavor than sweetness, and in those respects, these new varieties come up short.

Corn is a grain, but one that we eat in an immature state. If left on the stalk to full maturity, the kernels would become as hard as wheat and almost as full of starch. In fact, this is the state in which most of the corn grown in America is harvested — but those are different varieties that are processed for use in a whole range of industrial applications, including sweeteners, textiles and automobile fuels.

The kinds of corn we eat are picked within a month of pollination. In agriculture these varieties are referred to as "sweet corn." Because of their immaturity when picked, in the past they have also been called "green corn." (In the Southwest you still find green corn tamales, which are made with sweet corn rather than purely from masa, or ground dried corn.)

Almost every ear of sweet corn grown today was developed for a certain set of characteristics. This is not an example of modern-day Frankenfood genetic tinkering; it has been going on for centuries. (The ur-corn, teosinte, had cobs about two to three inches long that contained at most a half-dozen kernels.) But lately the march of progress has been especially swift. Besides traditional corn — which is practically nonexistent today (and would be so starchy you probably wouldn't like it even if you could find it) — corn breeders recognize three main families, or genotypes, of the new varieties, each with its own set of attributes. Agronomists refer to them by a two-letter shorthand.

The oldest improved variety is "su," for "sugary." This kind of sweet corn started being mentioned in seed journals in the 1820s but had probably existed before — the result of farmers selecting seed from the sweetest plants to propagate the next year. Most varieties of this type of corn have a sugar content ranging from 10 to 15 percent. That sugar starts converting to starch the moment the corn is picked. If left at room temperature, an ear of "su" corn will lose half of its sugar in less than a day. Even if chilled to normal refrigerator temperatures, it will lose two thirds of its sweetness within three days.

The next advance was a variety called "se," for "sugar-enhanced." This kind of corn is a *lot* sweeter than normal corn, containing as much as twice the sugar. The sugar-to-starch conversion occurs at about the same rate as for traditional corn, but because "se" corn starts out so much sweeter, it takes up to a week of storage before it falls to the sweetness level of normal corn.

The King Kongs of the corn world are varieties that are not only supersweet but that also go starchy much more slowly. These are called "sh2" corns because of the way their kernels shrivel and appear shrunken after drying. These varieties contain sugar levels between 30 and 45 percent — two to three times that of traditional corn. And their sugar-to-starch conversion rate is so slow as to be almost nonexistent. Even after being stored for a couple of days at warm room temperature (80 degrees), these varieties still have more than twice as much sugar as a freshly picked normal ear. They have been slow to win acceptance because the seed is significantly more expensive than that for the other improved corn types, and shoppers have been reluctant to pay the extra money.

Supersweet corn is the result of traditional plant breeding spurred by naturally occurring genetic mutations. Think of this breeding in terms of basketball players. In the general population, the occurrence of extremely tall humans is rare. But if two extremely tall people should find each other, fall in love and have children, the odds that their offspring will be extremely tall are, well, pretty short. And so it is with corn. Once breeders started working with a few "freak" corn plants that produced ears with very sweet kernels, it was just a matter of breeding and crossbreeding a few dozen generations to get where we are today.

The first real work on these supersweet corns was done by a University of Illinois professor named John Laughnan in the 1950s. But these varieties didn't catch on until the mid-1980s, and then they caught on quickly. In Florida, a prime winter corn-growing state, the percentage of supersweet corn went from 2 percent to 90 percent within five years. Today it is hard to find fresh corn grown anywhere that is not one of the improved supersweet varieties.

Understandably, these have become overwhelmingly popular among farmers and produce managers, who no longer have to listen to their customers complain about corn not being sweet enough. Unfortunately, that plaint seems to be giving way to another — that corn no longer tastes like corn. To an extent, that is true. Genetics is complicated, and it's hard to alter one factor without changing another. In the case of corn, increasing the sugar content has meant a decline in that amorphous quality called "corn flavor," as well as kernels that are no longer creamy (but crunchy) in texture.

What we think of as corn flavor is primarily based on aroma and is mainly a function of a chemical compound called dimethyl sulfide, which is also found in foodstuffs ranging from cabbage to lobster meat. (This chemical also poses a significant problem for brewers and vintners when it shows up in beer and wine.) Dimethyl sulfide is one of about half a dozen sulfurous compounds that appear in cooked corn (but not in raw corn, which is why raw corn always tastes so simple and sweet). It has a distinctive smell that is familiar from canned corn. Other sulfurous smells in corn are not so pleasant. For instance, the second-leading compound is hydrogen sulfide, which is familiar to anyone who has cracked a rotten egg. But put together in relatively low concentrations, these compounds add up to a lovely complex aroma.

Within the past couple of years, breeders have introduced varieties with complicated genetics that offer variations on the three basic genotypes. The goal is an ear of corn with the sweetness and slow sugar-to-starch conversion of "sh2," but with the creaminess and strong corn flavor of "su" and "se." Some of the new varieties try to accomplish this with a straight genetic blend — combining the best characteristics of each genotype in every kernel. Others take a different route — combining on the same cob kernels of each type of corn, so a single ear might contain 25 percent "su," 50 percent "se" and 25 percent "sh2." This last type is still scarce. It is expensive to grow, and so far, farmers say, their customers haven't been willing to put up with the additional expense.

Of course, when you're at the farm stand or produce market shopping for corn, odds are you won't have a clue as to whether the corn is "su," "se," "sh2" or any combination thereof. At best you'll be offered a choice of yellow or white — or bicolor, a cross-pollinated combination of the two. But despite what you may have been led to believe, one color of corn is not necessarily sweeter or "cornier" than another. The carotene that gives yellow corn its color is flavorless, and there are "su," "se" and "sh2" varieties of both white and yellow corn.

Really, then, it's all just packaging, and which color you prefer will depend to a great extent on where you live. Different areas of the country prefer different colors of corn. Generally speaking, white corn is preferred from the mid-Atlantic region through the South, bicolor is popular in the Northeast and yellow rules almost everywhere else. Most corn varieties have yellow kernels, so it could be that in some cases white varieties are preferred because they are less common and therefore in some way "special." Usually, though, preferences are determined by which color has traditionally been grown in the area. Sometimes, however, one special variety can influence buyers for generations. A study in Maine, for example, found that most of the state preferred bicolor corn because of the high quality of an old variety called Butter & Sugar — but this variety hasn't been grown commercially in the area for more than twenty years. In the southern part of the state, people preferred white corn because of a variety called Silver Queen, which is similarly antique.

Silver Queen, a very fine "su" variety introduced in 1955, has cast an inordinately long shadow in the sweet corn world. It has come to represent high quality, the "Cadillac of corn" as it were. This attraction exists despite the fact that blind taste tests have consistently shown that eaters prefer new "se" and "sh2" varieties. Perhaps more to the point, hardly anyone grows true Silver Queen commercially anymore, contrary to farm stand claims.

WHERE THEY'RE GROWN: Corn is one of the most widely grown vegetables, harvested in significant amounts in twenty-five states. More than half of the total U.S. production comes from just three states: Florida, California and New York.

HOW TO CHOOSE: The husks should be fresh and green with no drying. The silk should be golden and fresh-looking. Check out the tips of each ear: the kernels should be well filled out and evenly spaced. Pop a kernel with your thumbnail: it should spurt milky juice.

HOW TO STORE: Refrigerate corn, still in its husks, away from strong-flavored foods (corn absorbs odors). Keep it in its husks to help preserve the moisture in the kernels.

HOW TO PREPARE: Shuck corn right before cooking. Use a vegetable brush to remove the fine silk. If you want to cut the corn from the cob, place the cob tip down in a large bowl and slice downward along its length with a very sharp knife. You can then "milk" the corn to get any extra bits and juice by firmly stroking the length of the cob with the back of your knife (dull edge down).

ONE SIMPLE DISH: Everyone knows how to boil an ear of corn, but some people still don't know about grilling. It couldn't be easier. Soak whole, unhusked cobs in water for at least 20 minutes, then pop them on the grill. Grill the ears over a hot fire until they are a dark yellow and are well marked by the grill. This will take longer than you might expect — probably 25 to 30 minutes. Don't worry if the husks start to char; that just adds to the smoky flavor. Cool slightly and remove the husks and the silk before serving. (The fine filaments will come away more easily after grilling.)

Fresh Corn Blini with Crema Fresca

The combination of fresh corn, cornmeal and a smear of Mexican sour cream makes this an easy appetizer for an elegant summer dinner party. This recipe is loosely based on one in *Jeremiah Tower's New American Classics*.

8 SERVINGS, ABOUT 48 BLINI

- 1 cup yellow cornmeal
- 1½ teaspoons salt
- 1 teaspoon sugar
- 1½ cups boiling water
- 2 large eggs
- ¾ cup milk
- ½ cup all-purpose flour
- 2 tablespoons butter, melted, plus more for the pan
- 1 cup cooked corn kernels (1 ear)
- 1½ teaspoons minced red serrano or jalapeño chile
- Crema fresca (Mexican sour cream; available in Latin markets) or crème fraîche
- Fresh cilantro leaves

Combine the cornmeal, salt and sugar in a medium bowl and gradually stir in the boiling water to make a stiff paste. Let stand for 10 minutes to cool. Stir in the eggs one at a time, then the milk. Sift the flour over the mixture. Add the melted butter and stir until smooth. The texture should be somewhere between thick whipping cream and thin yogurt; if necessary, add a little more milk to thin it. Stir in the corn and chile. Cover and refrigerate for 30 minutes.

Heat the oven to warm. Heat a nonstick pan over medium-high heat. When it is hot, add a little butter and swirl it around the base of the pan. Pour in about 2 scant tablespoons of batter and cook until lightly browned and slightly crisp on one side, 2 to 3 minutes. You can cook 5 or 6 blini at a time.

When lightly browned on one side, turn and cook for 1 to 2 minutes on the other side. Transfer the cooked blini to a baking sheet lined with aluminum foil and keep warm in the oven. Repeat using the rest of the batter, adding more butter to the pan, if necessary.

Remove all the blini to a warm plate. Smear lightly with crema fresca and scatter cilantro leaves over the top. Serve immediately.

Grilled Corn and Arugula Salad

When you have leftover grilled corn, remember that all that smoky sweetness needs to be balanced with a fair measure of bite. In this recipe, that bite comes from the diced red onion and tart vinaigrette.

8 SERVINGS

- ½ garlic clove
- ½ pound red and yellow miniature tomatoes, cut in half
- ⅓ cup finely diced red onion
- 2 ears corn, grilled (see page 134) and cooled, or 2 ears leftover grilled corn
- ¼ cup olive oil
- 4 teaspoons fresh lemon juice
- ¾ teaspoon salt
- ½ pound arugula
- 1 ounce Parmigiano-Reggiano

Rub the inside of a large bowl with the cut garlic clove. Add the tomatoes and red onion. Using a sharp knife, cut the corn kernels cleanly from the cob into the bowl. Do not "milk" the cob to get the last bits (see page 134); that will muddy the appearance of the dish.

In a small bowl, whisk together the olive oil, lemon juice and salt. Place the arugula in a large bowl and toss it with enough of the dressing to coat lightly, about one third of the total amount. Arrange the arugula on a platter.

Add the remaining dressing to the corn and tomatoes and stir together gently. Spoon the corn mixture loosely over the arugula, then use a vegetable peeler to shave thin sheets of cheese over the top. Serve.

Shrimp and Sweet Corn "Risotto"

This isn't really a risotto, because it has no rice. But even the sweetest varieties of corn contain some starch, and that (and a little heavy cream) is what thickens the quickly made shrimp stock.

8 SERVINGS

- 1 pound shell-on medium shrimp
- ½ cup chopped green onions (trimmings reserved)
- ½ cup diced red bell pepper (trimmings reserved)
  Salt
- 4 small zucchini
- 6 ears corn
- 1 tablespoon butter
- 2 ounces Spanish chorizo or other mildly spicy dried sausage, diced (about 1 cup)
- ½ cup dry white wine
- ¼ cup heavy cream
- 3 tablespoons thinly sliced fresh basil leaves

Shell the shrimp, setting the shrimp aside and collecting the shells in a small saucepan with the green onion and bell pepper trimmings. Barely cover with water and bring to a simmer. Cook for 30 minutes, then remove from the heat and let steep for another 30 minutes. Season with salt to taste and strain into a measuring cup. You should have about 3 cups of shrimp stock. (Store any leftover stock, tightly covered, in the refrigerator for up to 1 week.)

Quarter the zucchini lengthwise, then slice crosswise about ⅜ inch thick. You should have about 4 cups.

Cut the corn from each cob. Holding each cob upright in a wide, shallow bowl, cut away the kernels with a sharp knife. Then, using the back of the knife (dull edge down), scrape any corn left on each cob into the bowl.

Melt the butter in a large skillet over medium heat. Add the chorizo and cook until well browned, about 10 minutes. Add the green onions and red bell pepper and cook until softened, about 5 minutes. Add the zucchini and shrimp and cook briefly.

Add 2 teaspoons salt and the wine. Raise the heat to high and cook until the wine is reduced to a syrup, about 5 minutes.

Add the corn and ½ cup of the shrimp stock and reduce the liquid until the pan is nearly dry. Add another ½ cup of the stock and the heavy cream and cook until slightly thickened and creamy, about 5 minutes. Remove the pan from the heat and stir in the basil. Taste for salt and serve.

Cucumbers

· · · · · · · · · · ·

When you get right down to it, cucumbers sometimes seem the very essence of "why bother?" All you get is a little crunch, a spurt of cool juice and — if you're lucky — a slightly herbaceous bitterness. Still, when the weather turns hot and humid, when the air is so thick that it sticks to your skin, a bite of something crisp and cool can seem like heaven. That's when we're thankful for cucumbers and all the work that has gone into raising them.

It took centuries of breeding just to make cucumbers edible. Wild cucumbers, which are found in the Himalayan foothills, are almost impossibly bitter. That someone once saw promise in this fruit is a tribute to the power of optimism (or extreme hunger). In fact, cucumbers are first cousins of the bitter melon, which is still appreciated in Southeast Asia. Both are members of the Cucurbit family, along with melons and winter and summer squash. As you can tell, this is one of the more varied families in all of horticulture, and that variety holds true with cucumbers as well.

Thousands of years of domestication have resulted in a bewildering variety of cucumbers. Not so long ago it seemed that you could find only one kind of cucumber — the familiar long, dark-green one, usually heavily coated with wax. But today you can find all kinds of cucumbers in the market, seemingly coming from every corner of the globe. There are Persian cucumbers, Chinese cucumbers, Armenian cucumbers, Japanese cucumbers, English cucumbers and Middle Eastern cucumbers. Some are more than a foot long, and others are only as big around as your pinkie. Some are warty; others are smooth. They come in every shade of green and even lemon yellow.

Although this range of variety is fascinating to horticulturists, it is less so to cooks. The one thing all cucumbers have in common is that they taste like cucumbers. Sure, there are very slight variations in degree of bitterness, and some cucumbers are a little crisper than others (although both of these traits are usually more attributable to farming than to genetics).

But other than that, cucumbers are all pretty much the same. Even the so-called lemon cucumbers get their name for the way they look — round and yellow when ripe — rather than the way they taste. The biggest difference between cucumbers, and the easiest way to differentiate among them for culinary purposes, is the thickness of the skin. Cucumbers with very thin skins have always been preferred for pickling (the salt can penetrate very quickly). That leaves cucumbers with thick skins for slicing. Of course, in the kitchen that doesn't make much sense at all, since thin-skinned cucumbers can be sliced just as easily as thick-skinned. In general, if you have to choose between cucumbers, pick whichever ones seem to be in the best condition rather than favoring a certain variety.

One of the great spurs in modern cucumber breeding has been the fruit's unfortunate effects on the digestive system. Put plainly, cucumbers sometimes make some people burp . . . a lot. At one time that burpiness was attributed to the seeds, and a lot of effort went into growing seedless cucumbers. These varieties have been developed to set fruit without being pollinated (or even produce almost entirely female flowers). They are cultivated in bee-free greenhouses (locked away in cucumber convents, as it were) to prevent the pollination that causes seeds. (Grown out-of-doors, these varieties will bear seeded fruit.) Unfortunately, although seedless cucumbers do have some culinary merit (the area surrounding the seeds in a cucumber is usually unpleasantly soft and watery), they don't help with indigestion.

Rather, the very bitterness that is so fundamental to cucumber flavor — and is one of the plant's more interesting traits — is what causes indigestion. The bitterness in cucumbers comes from compounds called cucurbitacins, which are normally found in high concentrations in the roots, stems and leaves of the cucumber plant rather than the fruit. Cucumbers are one of the more fragile plants a farmer can grow, susceptible to all kinds of fungi, viruses and insects, and these cucurbitacins serve as a kind of natural pesticide, discouraging many bugs from munching away. However, one bug, the cucumber beetle, has evolved with a resistance to the poisonous effects of cucurbitacins. In fact, it is downright fond of the flavor. Not only do cucumber beetles love to eat cucumber plants, but they also harbor harmful bacteria that can cause severe wilt, which can kill the plants. So much for defense mechanisms.

In most cases, cucurbitacins are not found in the fruit in sufficient quantities to spoil the taste. But when cucumber plants are stressed — when the weather turns suddenly hot and dry — they tend to produce fruit that is more bitter. (Perhaps, sensing a threat, the plant is trying to protect its seeds.) Some varieties are less susceptible to stress than others — generally look for anything with "sweet" or "burpless" in the name. Unfortunately, these also taste pretty bland.

Fortunately, you can choose more flavorful varieties and take care of much of the burpiness in the kitchen. Cucurbitacins tend to be concentrated just under the skin and around the stem of the cucumber. So you can reduce (if not eliminate) burpiness by peeling deeply and removing the stem end. This step will be sufficient for most people; however, among those who are extremely susceptible to burpiness, any cucumber will have some effect.

Although they may not look it, cucumbers are very sensitive fruit. More than 95 percent water, they begin to soften almost as soon as they are picked. Without treatment, they should be eaten within three or four days. The most common method used for extending the shelf life of cucumbers is coating them with wax to slow the loss of moisture. This wax is technically edible, but it has an unpleasant greasy quality even when washed. If you can find only waxed cucumbers, peel them.

Another alternative shippers use is sealing cucumbers tightly in plastic. This is more expensive, but it does not leave any residue on the fruit. Water loss is not the only threat to cucumbers. They are also susceptible to chilling damage. If refrigerated for very long, they'll begin to develop pitting on the surface. Keep them tightly wrapped in the warmest part of the refrigerator (about 50 degrees is ideal) for the best results. Cool as a cucumber, yes, but not too cold.

WHERE THEY'RE GROWN: Georgia grows about a quarter of all the cucumbers in the United States, followed closely by Florida. About 45 percent of the cucumbers eaten in the United States are imported, primarily from Mexico.

HOW TO CHOOSE: Remember that the worst thing that can happen to a cucumber is moisture loss, so avoid any that look shriveled or wilted.

HOW TO STORE: Seal cucumbers tightly in a plastic bag and keep them in the refrigerator. Use them quickly.

HOW TO PREPARE: The only cucumbers that need peeling are those with very thick skins and those that have been waxed. All others can be sliced with the skin on. Some cucumbers have large seed cavities in the center. To remove the seeds, simply carve them out with the tip of a teaspoon.

ONE SIMPLE DISH: A salad of cucumbers and yogurt is so simple and perfect that it is served all around the world. Slice cucumbers into a bowl and add just enough yogurt to dress them lightly. Stir in salt and some fresh herbs — dill is a natural, but basil is good, too. This salad is best prepared just before serving so the cucumbers don't release too much moisture into the dressing, watering it down.

Cucumber Gazpacho

This is a variation on the familiar combination of cucumbers and yogurt (see page 143). Using stale bread as a thickener adds a silky texture, and the sorrel leaves underscore the refreshingly tart flavor. The dark, thin-skinned cucumbers called Persians are the best ones for this recipe. You can use other thin-skinned cucumbers, but the color won't be as pretty. If you use regular slicing cucumbers, peel them and remove the seeds.

6 SERVINGS

- 8 ½-inch-thick slices stale baguette
- 2 pounds cucumbers
- 1½ ounces fresh sorrel leaves, stems removed
- 1½ teaspoons minced garlic
  Salt
- 4 cups low-fat plain yogurt, plus more for garnish

Tear the baguette into rough pieces and put it in a bowl with water to cover. Soak for at least 15 minutes.

Coarsely chop the cucumbers and place them in a blender. Chop most of the sorrel leaves, reserving 2 for garnish. Add the sorrel leaves to the blender along with the garlic, 1 teaspoon salt and yogurt and puree until smooth.

Remove the bread from the water and squeeze dry. Add the bread to the blender and puree until perfectly smooth. Pour the mixture through a strainer into a deep bowl, discarding any bits of bread caught in the strainer. The soup should be slightly thickened, about the texture of heavy cream. Cover the bowl tightly and refrigerate for at least 1 hour.

To serve, season the soup with more salt to taste and ladle it into wide bowls. Use a large spoon to swirl in a streak of yogurt. Thinly slice the reserved sorrel leaves and scatter a few slices across the top of each bowl.

Cucumber, Beet and Feta Salad

The thing that makes this salad so special, besides the delicious interplay of flavors, is the textures. The beets are almost buttery smooth, the cheese is crumbly and the cucumbers set everything off by providing a pleasing crunch.

4 SERVINGS

- 1 **pound beets, tops trimmed (about 4)**
- 1½ **pounds cucumbers, unpeeled (about 2)**
- 3 **tablespoons olive oil**
- 1 **tablespoon sherry vinegar**
- 2 **teaspoons minced mixed fresh herbs (preferably chives and tarragon)**
- 1 **teaspoon minced garlic**
- ½ **teaspoon salt**
- ¼ **pound feta, crumbled**

Place the beets in a large saucepan with plenty of water to cover. Bring to a boil and cook at a fast simmer until the beets can be pierced easily with a sharp knife, about 45 minutes. Drain, rinse under cold running water and shake off any excess moisture.

Cut away the ends of the cucumbers and slice them a little less than ¼ inch thick.

In a large bowl, whisk together the oil, vinegar, herbs, garlic and salt, mixing well. Add the cucumbers and stir to coat well. Remove the cucumbers with a slotted spoon, draining any excess dressing back into the bowl, and arrange on a platter in a broad layer.

Using your fingers, slip the skins off the beets. Slice the beets ¼ inch thick, add them to the leftover dressing and stir to coat well. Remove the beets with a slotted spoon and arrange them in an oblong mound on top of the cucumbers, centering the mound so the cucumbers show around the edge.

Sprinkle the feta over the top and serve.

Eggplants

An eggplant is a thing of rare beauty. Its form ranges from as blocky and solid as a Botero sculpture to as sinuous and flowing as a Modigliani. Its color can be the violet of a particularly magnificent sunrise or as black as a starless night. It can be alabaster white or even red-orange and ruffled. And its beauty is more than skin-deep. The flesh is at once luxurious in texture and accommodating in flavor. So why does the eggplant scare people?

Most of it has to do, in one way or another, with the vegetable's supposed bitterness. Of course, there's the classic recommendation that eggplant needs to be salted before cooking to remove the bitterness. But that's just the start. Eggplants with large calyxes (the leafy-looking green part that connects the vegetable to the stem) are bitter. Eggplants with more seeds are bitter. Eggplants that are heavier are bitter. At least that's what some people say. Others claim just the opposite: it's the lighter eggplants that are bitter. Some of the assertions take an almost psychological turn: Eggplants that are old are bitter. Eggplants with darker skins are bitter. Eggplants that are male are bitter. (For the record, botanically speaking eggplants are fruits and therefore neither male nor female.)

Let's get one thing straight: most eggplants are not bitter (even though they have every right to be after everything that has been said about them). At least they are no more bitter than a green bell pepper or the tannic skin of a fresh walnut. They have a whisper of bitterness that adds to the taste rather than ruining it. In fact, it's that subtle edge that makes eggplant such a great companion to so many flavors. Without that edge, it would be bland, nothing more than field-grown tofu. But that earthy undertone serves to focus our attention on other flavors, the way a bass line complements a melody.

Combine that natural accommodation with a spongelike absorbency, and eggplant is one of nature's great sidemen. It soaks up the flavor of whatever it is cooked with, and somehow amplifies and smoothes out that flavor in the process. Good olive oil has no greater friend than eggplant and vice versa. Fry eggplant in olive oil, and what once was a hard, dry, almost pithy vegetable becomes downright voluptuous. The surface crisps slightly, and the inside turns creamy and smooth. (Also make sure to brush eggplant with oil before grilling. It will keep the surface from drying out.)

The secret to great fried eggplant is actually one of the supposed cures for bitterness. Salting the vegetable does nothing to remove bitterness (which isn't really there in the first place), but it does pull the water out of the eggplant, collapsing the cells, which then absorb oil more easily during cooking. Try it and you'll see. Cook salted and unsalted eggplant side by side with oil in a skillet and dry on the grill. Salting will make absolutely no difference in the grilled eggplant, but it will in the fried. Unsalted fried eggplant is meaty; salted is creamy. It all depends on what you like. Don't shortcut this step. It takes about an hour of purging (an hour and a half is better) to make a difference. Some cooks recommend pressing the eggplant under a weight during this process. Although this makes sense in theory, I found that pressing resulted in eggplant disks that cooked up like wafers rather than pillows.

Some cooks also claim that salting reduces the amount of oil the eggplant absorbs during frying. This, unfortunately, is not true. Salted and unsalted both soak up prodigious amounts of oil — as much as 2 tablespoons per ½-inch-thick slice. Supposedly, if you cook eggplant longer, it will release the oil it has absorbed as the cell structure collapses. This takes very slow, patient cooking, however, over a long period of time.

One thing that might rightly intimidate people about eggplants is the sheer variety that's available. What is a cook to make of a vegetable that can take so many different forms? There are so many eggplants in the world that it's impossible to keep up with them. In fact, scientists aren't even sure of the exact number. (In his *Cornucopia II*, an authoritative guide to edible plant life, Stephen Facciola lists fifty-six major eggplant varieties.) Beyond its ancestral home in Burma, eggplant is a staple food in India, China, Southeast Asia, much of Africa and the Mediterranean. And as is so often the case after centuries of small-scale subsistence cultivation — where farmers save seeds from year to year, gradually changing the plant's genetics — the varieties are poorly defined, with one type shading into another.

Some of these varieties look so unusual that you wouldn't even know they were eggplants. The elaborately tufted, lavender-skinned Rosa Bianca is as big as a toddler's head, and the beautifully marbleized green Thai eggplant is smaller than a golf ball. Creamy oval eggplants about 3 inches tall look just like eggs. There are long, thin eggplants that range in hues from green to black-purple to violet to white. Tiny Thai "pea" eggplants look for all the world like very small green peas that grow in clusters like grapes. (Discussion continues among botanists about whether this is a true eggplant or a close cousin.)

It would be nice to say that the visual variety of eggplants is matched by an equally wide range of flavors, but that would be a lie. For the most part, eggplant tastes like eggplant. This is not to say that all eggplants are interchangeable. Eggplants vary in how thick their skin is and how seedy they are, and they vary in the exact texture of their flesh. But they don't vary much in flavor.

So the little green Thai eggplant, although it is very seedy and crunchy, tastes pretty much like the small, thin Chinese "finger" eggplant, which has very few seeds, creamy flesh and an amazingly thin skin. And that in turn tastes like the familiar blocky black eggplant, with its thick skin, coarse flesh and moderate amount of seeds. Furthermore, except for the blackest of eggplants, the skin color fades during cooking, resulting in a muted palette of shades of greenish beige.

WHERE THEY'RE GROWN: Eggplants are a pretty minor crop in the United States. Production is fairly evenly split among Florida, California, Georgia, New York and North Carolina. In fact, Mexico grows more of our eggplants than all of the United States combined.

HOW TO CHOOSE: For a vegetable that can look like such a brute, eggplant is surprisingly fragile. It bruises easily, and those bruises quickly turn bad. (Cut open a dented eggplant, and you'll see that the flesh is brown and corky in the affected area.) It also loses moisture quickly, leading to dry and pithy flesh. When choosing an eggplant, pick one that is heavy for its size; that will be the freshest. Also feel the skin. If it is a round eggplant, the skin should be taut and almost bulging. The long, thin eggplants found in Asian markets are often slightly softer, but they should not be so soft that the skin is wrinkling.

HOW TO STORE: The eggplant is a tropical plant and hates the cold. Bronze patches on an eggplant's skin are signs of chill damage, which can occur after the fruit is picked as well as before. In an ideal world, you'd buy only enough eggplant to use for one day, and you'd store it in a cool spot on the counter. (Eggplants hate to get colder than 45 degrees, and most home refrigerators are between 35 and 40 degrees.) Eggplants' thin skin is also susceptible to water damage, so keep your eggplants as dry as possible. The best solution is to store them in the crisper drawer of the refrigerator, in a plastic bag with a crumpled-up sheet of paper towel to absorb excess moisture. Kept this way, they'll be of acceptable quality for up to a week.

HOW TO PREPARE: Eggplants can be peeled or not, depending on your preference. The peel is slightly tough, and if you're cooking an eggplant whole, the peel can split during the process. You can also peel the eggplant in alternating lengthwise strips, which gives it a pretty harlequin effect. Salt eggplant only if you are going to fry it in oil.

▶ ONE SIMPLE DISH: To grill eggplant, cut it lengthwise into ½-inch-thick slices. Brush both sides of each slice with garlic-flavored olive oil. Continue brushing lightly during cooking. Grill just until the eggplant is tender — try not to char it, although that's unavoidable to a certain extent. When a slice of eggplant is done, transfer it to a serving platter, then sprinkle it with salt and minced fresh herbs. When the next slice is done, place it on top of the first, repeating the seasoning. Continue layering the slices until all are cooked.

Smoky Eggplant Bruschetta

Baking an eggplant brings out a somewhat unexpected smoky character (think of baba ghanoush). Be sure to pierce the eggplant before cooking, or it might explode. Once you've peeled the baked eggplant, all it needs is a rough stirring to turn it into a coarse puree. Resist the urge to stick it in the food processor. This not only dirties extra dishes but also gives the eggplant the texture of baby food.

6 SERVINGS

- **2 1-pound eggplants**
- **2 teaspoons minced garlic**
- **1 teaspoon minced fresh rosemary**
- **1 tablespoon fruity olive oil**
- **Salt**
- **2 teaspoons red wine vinegar**
- **1 teaspoon fresh lemon juice**
- **1 tomato, diced**
- **1 baguette**
- **About 2 ounces pecorino Romano**

Heat the oven to 400 degrees. Pierce the eggplants in 2 or 3 places with a sharp knife and place them on a jelly-roll pan or in a baking dish. Bake until the flesh is soft and the eggplants have collapsed, about 1 hour. Remove from the oven and let cool.

When the eggplants are cool enough to handle, peel away the skin and coarsely chop the flesh. Put the flesh in a bowl with the garlic, rosemary and olive oil and stir roughly with a wooden spoon so that the eggplant shreds and breaks apart into chunks but does not become a smooth puree. Stir in 1 teaspoon salt, the vinegar and lemon juice. Taste and adjust the salt. Gently stir in the tomato.

Slice the baguette about ½ inch thick. Toast the bread slices under the broiler or on the grill until browned on both sides. Spoon some of the eggplant mixture on each slice and use a vegetable peeler to shave a thin slice of pecorino Romano on top. Serve at room temperature.

Silken Eggplant Salad

Steamed eggplant? It may not be the most common treatment, but it is the best way to get to the elusive pure, sweet but earthy flavor of the vegetable. It is surprisingly delicious. When this was served to a crowd of non–eggplant enthusiasts at a dinner, it disappeared before the tomato salad.

6 SERVINGS

- 2 **pounds long, thin eggplants**
- 1 **tablespoon minced garlic**
- ½ **cup olive oil**
- **Fresh lemon juice**
- **Salt**
- ½ **teaspoon minced fresh rosemary**
- ¼ **teaspoon red pepper flakes**
- **Generous grinding of pepper**

Remove the tough green calyxes from the eggplants and cut into thirds or halves to fit in your steamer basket. Steam over rapidly boiling water until tender, about 5 minutes. The perfect point of doneness is when a knife slips easily into the eggplant but the skin doesn't wrinkle (this happens when the flesh begins to pull away from the skin).

Remove the eggplant from the heat and set aside until it's just cool enough to handle. The hotter the eggplant is at this point, the better it will absorb the dressing.

While the eggplant is cooling, whisk together the garlic, olive oil, 2 tablespoons lemon juice, 2 teaspoons salt, the rosemary and red pepper flakes in a bowl.

Cut the eggplant pieces in half lengthwise. If the bottom ends are very thick, cut these into quarters. Cut the eggplant into 2-inch pieces and add them to the dressing. Toss well to combine, then cover and refrigerate until ready to serve.

To serve, add the pepper, then taste the eggplant and adjust the salt, pepper and lemon juice.

Grilled Eggplant with Walnut-Cilantro Pesto

After you've had your fill of grilled eggplant dressed simply with olive oil and herbs (see page 150), there's no need to pack up the charcoal. This suave sauce made from pounded walnuts and cilantro makes for an elegant dish with a haunting flavor that is vaguely Turkish.

6 SERVINGS

½ teaspoon red pepper flakes
2 garlic cloves, minced
   About 1¼ cups olive oil
¼ pound shelled walnuts
   Salt
1 teaspoon paprika
3 tablespoons chopped fresh cilantro
2 tablespoons fresh lemon juice
1½ pounds long, thin eggplants

In a bowl, combine the red pepper flakes, half the garlic and ½ cup of the olive oil. Let stand for at least 5 minutes.

Using a mortar and pestle, crush the remaining garlic, the walnuts and ¼ teaspoon salt to a paste. Add the paprika and then slowly add about ¾ cup olive oil, stirring it in with the pestle, until the sauce has the consistency of loose, coarsely ground peanut butter. Stir in the cilantro, grinding it with the pestle to release its oils. Stir in the lemon juice; taste and adjust the salt and lemon juice. Be particularly careful with the salt. The sauce should not be overtly salty, but I usually have to add a little more to get the full flavor of the walnuts. Set aside until ready to use.

Slice the eggplants lengthwise about ½ inch thick, leaving the slices attached at the stem end. Brush the cut surfaces with the flavored olive oil and grill over medium heat. Turn the eggplants frequently so that they cook through without scorching. Before each turn, brush lightly with the olive oil. You'll know that the eggplants are cooked through when the slices flop lazily when you turn them.

Arrange the eggplants on a platter, spreading the cut slices slightly. Spoon half of the walnut pesto over the top and serve, passing the remaining pesto at the table.

Green Beans

· · · · · · · · · · · · ·

How do you keep green beans green? Granted, it may not be the most pressing question of our time, but it is one that cooks everywhere scramble to find an answer to every summer. Although the method is surprisingly simple, it turns out that sometimes green beans shouldn't be all that green anyway.

Every green vegetable goes through essentially three stages of green. When it is raw, the vegetable is a deep but dull green. During the early stages of cooking, the color turns bright and vibrant. This change occurs as the cell walls soften and the tiny amounts of oxygen and other gases they contain, which cloud the pure color of the chlorophyll, are driven off. Finally, the green turns olive drab. This happens because of a chemical change in the chlorophyll, which is partly due to an enzymatic action, but mostly due to acids that either are introduced during cooking (adding vinegar or tomatoes, for example) or come from the vegetable itself and are released during cooking.

Since vibrant color is so important, cooks have devised different strategies for getting around this chemical change. Old French chefs added a bit of baking soda to the water. This helps preserve the green, but it also speeds the breakdown of the plant's cellulose, making the texture slimy. Some modern chefs insist that cooking in very heavily salted water will help. There doesn't seem to be any scientific basis for this claim, although it probably does no harm and will certainly help the seasoning of the vegetables. The best way to avoid olive drab vegetables is to cook them in plenty of water (to dilute the acids that are released); to make sure the water returns to the boil as quickly as possible (this will finish the cooking more quickly); and to cook them briefly — then, if possible, plunge them into ice water as soon as they are done (this stops the cooking process immediately).

Although a vibrant green color is usually desirable, there are times when it can be forgone in favor of flavor. In the case of green beans, for example, the importance of preserving the color depends largely on what kind of bean you're cooking.

The green bean is a legume that is harvested when immature. If green beans are left to grow to maturity, the seeds inside will swell to nugget size, and the tender pods will toughen and become coarse (as any gardener who has ever gone on summer vacation will readily attest).

Green beans can be roughly divided into two groups: the round and the flat. The round types, most famously Blue Lake and the French haricots verts, tend to be more delicate in texture and brighter in flavor than the flat types. These should almost always be cooked briefly to preserve their color. Flat beans, such as the Italian Romano, are meatier and more assertive. You can cook them longer than round beans — long enough even that the color will fade to olive drab. In fact, because of their thick, meaty texture, their flavor actually seems to improve with long cooking.

"Green" beans also come in different colors. There are chlorophyll-free yellow beans and several varieties that are purple. And there are beans that are purple and yellow, such as the dragon tongue bean. Yellow beans keep their color during cooking, but purple and patterned beans fade to green — a dull green at that. The Asian yard-long bean, which looks like a stretched-out green bean (it comes in purple as well), is actually more closely related to the black-eyed pea. Its flavor reflects that lineage, but its colors are bound by the same scientific laws as regular green beans.

Green beans are sometimes called "string beans," although that is a misnomer — or at least an antique usage. A tough filament runs up the seam of old green bean varieties and needs to be removed before cooking (just as with edible pod peas). The string was bred out of most popular varieties around the turn of the century, but you can occasionally find heirlooms that still have it. The old name "snap bean" is much more to the point and still valid, as green beans wilt quickly and should be crisp enough when you buy them that they'll snap cleanly in half.

Some legumes are left to grow to full maturity, whereupon they are shelled and dried for use through the winter. Dried beans come in sizes ranging from a grain of rice to as big as your thumbnail; in a wide assortment of jewel-like colors — reds, pinks, purples, maroons, even black; and in an almost countless array of patterns. (Unfortunately, for the most part these brilliant colors fade to various shades of tan and brown during cooking, although some of the patterns are distinct enough still to be visible.) Dried beans are picked fully mature, after most of the sugars have turned to starch. The drying after harvest removes all but the last traces of moisture, so they can be stored almost forever.

That also means that dried beans require extensive cooking in plenty of water to become edible, as those rock-hard starch granules need not only to soften but also to reabsorb the lost moisture. You can shorten the cooking time somewhat by soaking the beans beforehand (though, contrary to popular opinion, this does little to allay dried beans' famous gastric side effects, and soaked beans never come close to the richness of beans that have been cooked without soaking). Depending on the size of the bean and how long it's been stored, soaking can save as much as an hour of cooking. But soaking also means that you have to plan dinner the night before. You're really just choosing between inconveniences.

Some beans are harvested at an intermediate stage. Shelly beans, which are between green and dried, offer some of the most sublime eating you can imagine, but they are difficult to find except in the South or at farmers' markets. They are the same varieties that are normally dried but are sold at full moisture. They cook in 20 to 30 minutes without any special treatment and have a silkiness of texture and a sweetness and delicacy of flavor that are unmatched. Probably the stellar example of this is the lima bean. Scorned by many who know it only in its dried, canned or frozen form, the lima at the shelly stage is nothing short of spectacular, with a complexity of flavor that puts even favas to shame.

WHERE THEY'RE GROWN: Only about a quarter of the green beans grown in the United States are eaten fresh, and almost half of those come from Florida. Georgia and California trail far behind. Imports represent about 10 percent of the beans sold, but that percentage is increasing, particularly during the winter months. Mexico supplies more than 90 percent of the imported fresh beans.

HOW TO CHOOSE: Green beans are surprisingly delicate and begin to lose moisture as soon as they are picked. The pods should have no sign of wilting or mold. The name "snap bean" is a great clue: bean pods should be so crisp that they snap when bent.

HOW TO STORE: Green beans are somewhat hardier than peas and fava beans, but they still need to be cooked soon after you buy them. Store them tightly wrapped in the crisper drawer of the refrigerator.

HOW TO PREPARE: Most green beans no longer have strings, but they do have stems and tough stem ends, which must be removed. The trailing point at the other tip of the bean can be either cut off or left on.

ONE SIMPLE DISH: There is no easier or more delicious accompaniment to any grilled or roasted meat than a bowl of green beans. The best way to prepare them is to blanch them in a big pot of rapidly boiling salted water for about 7 minutes. They should be bright green and al dente — well cooked, but with a slight crispness. When you pick a bean out of the water, it should droop but not sag. Stop the cooking in ice water and pat the beans dry. Dress them with a mixture of olive oil, lemon juice and just a bit of garlic right before serving. (If you dress them too early, the lemon juice will change the color.)

Chicken Salad with Green Beans and Basil Mayonnaise

During late spring and early summer, this is the kind of meal we live on at my house. The best beans to use are the pencil-thin ones, such as Blue Lakes. The fatter Romanos are just a little too tough for this dish. And the painfully skinny haricots verts don't have enough oomph to hold their own.

6 SERVINGS

- 1 cup mayonnaise
- ¼ cup thinly sliced loosely packed fresh basil leaves
- 2 teaspoons herb vinegar
- ½ pound watercress, tough stems removed and torn into bite-size pieces
- 1 pound cooked green beans, cut into bite-size pieces
- ¾ pound cooked skinless, boneless chicken, preferably grilled or smoked, cut into bite-size pieces
- ¾ pound cherry tomatoes, cut in half

In a medium bowl, beat together the mayonnaise, basil and vinegar until smooth.

In a large bowl, combine the watercress and 2 tablespoons of the basil mayonnaise and toss until the leaves are lightly coated. Divide among six plates.

To the same bowl, add the beans, chicken and tomatoes and ¼ cup of the mayonnaise. Stir to combine well, making sure all the pieces are very lightly coated. Divide evenly and place atop the watercress.

Top each salad with a teaspoon or so of dressing. Serve, passing the remaining dressing at the table.

Overcooked Green Beans

Meaty Romano-type green beans can take extended cooking — in fact, they even reward it, turning silky and rich-flavored. Don't worry if the color turns drab: the benefits far outweigh any lack of brightness.

4 SERVINGS

- 2 ounces dried salami, diced (about ¼ cup)
- 2 tablespoons olive oil
- 1 onion, chopped (about ⅔ cup)
- 2 garlic cloves, minced (about 1 tablespoon)
- 2 plum tomatoes, peeled, seeded and diced (about 1 cup)
- Salt
- 1 pound flat (Romano-type) green beans, stem ends snapped off
- ½ cup water
- ½ teaspoon balsamic or sherry vinegar

Warm the salami in the oil in a large skillet over medium heat. Add the onion and cook until fragrant, about 3 minutes. Add the garlic and cook for another 3 minutes. Add the tomatoes and cook until they melt into the sauce, about 5 minutes. Season to taste with salt.

Add the green beans and the water, cover tightly and shake the pan to coat the beans with the sauce. Reduce the heat to low and cook until the beans are extremely tender and flavorful, about 30 minutes.

Just before serving, stir in the vinegar and season to taste with salt. The beans can be served either hot or at room temperature.

Summer Squash

. . . .

Surely, zucchini must have been with us always. It is one of the most widely grown crops in the world, with substantial harvests in North and South America, Europe, India, the Middle East and Asia. More than 6 million tons of summer squash are grown every year around the world, with zucchini accounting for the lion's share in almost every location. Sometimes it seems that much is grown in each of our neighborhoods alone. Every summer we struggle to come up with new ways to cook it, fighting to keep our heads above what seems to be a rising tide of squash. We stuff it, stew it, sauté it, steam it and serve it raw in salads. And still we never seem to make a dent.

So it may come as a surprise to learn that zucchini, which seems so ancient, is actually a relatively modern invention, probably dating back no earlier than the turn of the twentieth century. According to squash historian Harry Paris, the first recorded mention of zucchini came in a 1900 Italian seed catalog. It probably didn't make its way to the United States until after World War I, brought by Italian immigrants to California.

Why the doubt? How hard can it be to spot a zucchini? Actually, it can be pretty danged difficult. Certainly, there were cylindrical green summer squash harvested much earlier than the twentieth century. A basket of them, flowers attached, appears in *Fruttivendola,* a well-known sixteenth-century painting of a fruit and vegetable merchant by Vincenzo Campi. But those aren't zucchini. They are cocozelle, one of zucchini's forebears. There are also records of cylindrical green squash being harvested throughout the eastern Mediterranean since fairly shortly after Columbus's voyages. These are not zucchini either, but marrow squash. To make matters even more confusing, all three of these are just broad families; within each are dozens, if not hundreds, of individual varieties. And those are just the cylindrical green ones. Summer squash also come in other colors and shapes. At the Newe Ya'ar Research Center in Israel, Paris grows more than 320 varieties he has deemed of special interest.

Making sense of all this squash is not nearly as complicated as it may seem. Start with the cylindrical green ones, since they are now the most popular. Zucchini are dark green and show little or no taper along their length (they are roughly the same circumference at the flower end as at the stem end).

Marrow squash are generally pale to grayish green in color and tapered in shape (they are bigger around at the flower end). Still preferred in the eastern Mediterranean and Latin America, they tend to have denser flesh than zucchini and so hold their shape better in soups and stews.

Cocozelle, which are mostly grown in Italy, are longer and thinner than zucchini and slightly bulbous at the flower end. Whereas a zucchini is usually 3½ to 4½ times as long as it is wide, cocozelle can be 8 or more times as long. The distinctions between the two are more than in shape and color, though. Cocozelle tend to have a richer flavor than zucchini.

But, of course, nothing is quite that simple. Squash are notoriously promiscuous. More than one hundred specific varieties of zucchini are grown today, and many of them are crosses between true zucchini and either marrow squash or cocozelle (sometimes both). Knowing this background makes it easier to guess the qualities of the squash you see. If it is dark green and extremely thin for its length, it probably has some cocozelle lineage and a fairly strong flavor. If it is pale green and slightly bulbous, it is more likely to have denser flesh.

In addition to these cylindrical green varieties, three other families of summer squash are popular. Crookneck squash are yellow, with narrow, bent necks and bodies that become quite bulbous toward the blossom end. Straightneck squash are similar, except the necks are not bent. (Just as all cylindrical green squash are sometimes lumped into one family, so are the crooknecks and straightnecks.) Scalloped squash, which can be either yellow or green, are flattened with scalloped edges.

All three families are mildly squashy and less richly flavored than good zucchini. Still, until the advent of the zucchini, they were the most popular summer squash grown in the United States, and quite a bit of regional allegiance to them lingers. Crooknecks are especially popular in the Southeast, while straightnecks are preferred in the Northeast. Scallops, which once had pretty universal popularity, are beginning to make a commercial comeback after having fallen from favor — though not in the range of varieties that were once available.

The relatively new Golden Zucchini, released in 1973, is also becoming popular. And at farmers' markets you may see round zucchini, such as the several Italian varieties called Tondo and the Provençal Ronde de Nice. These are not true zucchini, but can be more accurately described as "summer pumpkins." They have firm flesh and mild flavor similar to a marrow squash.

All of these summer squash are harvested at various stages of immaturity, usually within a week of their flowers becoming fully open. Left to grow indefinitely, they would become almost as big and hard-shelled as their pumpkin cousins. But whereas pumpkins develop dense, sweet flesh when fully grown, summer squash get spongy and watery. This is particularly true of crooknecks, which even when fairly small can have hard, warty skin and a mushy interior. Zucchini are usually harvested when four to five inches long, though sometimes they are picked at little more than an inch (and in this case frequently with flowers still attached). Up to six or seven inches long, zucchini are good for stuffing, as the spongy center is removed. Much bigger than that, and the squash quickly becomes bitter with fully developed seeds and pith — the stuff of zucchini jokes.

Unlike the flowers of most vegetables, squash flowers can be eaten as well as the fruit. They have a delicate flavor reminiscent of the squash itself. Dip them in a simple flour and water batter and deep-fry them either as they are or stuffed with a little meat or cheese. Or chop them up and stir them into a risotto, a frittata or a stew for stuffing a quesadilla. This is most commonly done with zucchini flowers, but other squash flowers, including pumpkin flowers, can be used as well. These flowers can be either female (with the fruit attached) or male (no fruit, intended only to produce pollen). A single plant will produce flowers of both types. Because they are so delicate, squash blossoms must be used the day they are picked and so are rarely found in groceries. If you want them, you'll have to cultivate a farmers' market grower or your neighborhood gardener. If the grower is reluctant, just remind him that the more flowers he sells, the fewer zucchini he'll have to worry about.

WHERE THEY'RE GROWN: Despite their name, summer squash are actually mild-weather crops and are available on a fairly constant basis throughout the year. Three states lead the market — Florida, California and Georgia — but it is a slim lead indeed; the top five states account for less than 55 percent of the squash grown. Squash sales are booming — consumption has increased by almost a third in the past decade. To keep up with this demand, imports — almost entirely from Mexico — have become an important part of the supply picture, accounting for almost 40 percent of the squash consumed in the United States (530 million pounds imported vs. 885 million pounds domestic).

HOW TO CHOOSE: Whatever the type, summer squash should be firm and free of wrinkles and nicks. Really fresh squash will bristle with tiny hairs. The zucchini and its relatives and scalloped squash should be deeply colored, but crookneck and straightneck squash should be pale — by the time their rinds have turned gold, they will be hard and warty (this is not true of Golden Zucchini).

HOW TO STORE: Summer squash are fairly perishable and should be cooked within a week of harvest. They should be refrigerated until ready to use, sealed in plastic bags to slow respiration. Do not wash the squash until right before cooking, as any moisture on the skin will make it spoil faster.

HOW TO PREPARE: Summer squash are eaten whole, peel and all, so the only preparation necessary is removing the remains of the stem and any scar at the blossom end. Since squash cook fairly quickly, it makes a big difference how thick it is cut. Sliced thin, it will melt away to a rough puree. If you want pieces to remain intact, cut it into thicker sections.

ONE SIMPLE DISH: Summer squash is the perfect vegetable for braising. Cut it up and cook it with a little olive oil, about 2 tablespoons water and some garlic in a covered skillet over medium-high heat. When the squash begins to become tender, remove the lid and increase the heat to high. Cook, stirring constantly, until the squash begins to glaze.

Zucchini Frittata

Serve wedges of this frittata with a salad to make a summer supper, or cut it into cubes and spear them with toothpicks to use as an appetizer.

6 MAIN SERVINGS

 1 **pound zucchini**
    Salt
 ½ **onion, sliced**
 3 **tablespoons minced fresh parsley**
 2 **tablespoons olive oil**
 6 **large eggs**
 2 **ounces Parmigiano-Reggiano**

Shred the zucchini using a coarse grater. Salt liberally (using 2 to 3 teaspoons), stir well and place in a colander to drain for 30 minutes.

Heat the broiler. Combine the onion, parsley and olive oil in a medium nonstick skillet over medium heat. Cook until the onion softens slightly but does not begin to brown, about 5 minutes.

Stir the zucchini in the colander to get rid of as much excess liquid as possible. Add the zucchini to the skillet and cook briefly, stirring to mix well. Reduce the heat to medium-low.

Beat the eggs with a fork in a medium bowl just until they are of a uniform consistency. Add the eggs to the skillet and stir to combine with the vegetables. Cook until the surface just begins to set, about 10 minutes.

Grate the cheese over the top, distributing it as evenly as possible. Place the skillet under the broiler, as far from the flame as possible, and cook until the top is set and the cheese just begins to brown, 3 to 5 minutes.

Let stand at room temperature for 5 minutes. The frittata should be just beginning to pull away from the sides of the skillet. Run a thin spatula under the frittata and shake the pan to free any sticking spots. Slide the frittata cheese side up onto a plate. Cut into wedges and serve.

Garlic-and-Herb-Stuffed Tomatoes and Zucchini

A simple way to stuff a zucchini, this method has the distinct advantage of showing off the squash's flavor. Baking the squash with small, round tomatoes stuffed with the same mixture adds enough complexity that you can serve this dish as a light summer dinner.

6 SERVINGS

- ⅓ cup pine nuts, toasted
- 2 tablespoons olive oil
- 1 onion, minced
- 4 garlic cloves, minced
- 1 28-ounce can crushed tomatoes
- ½ cup dry white wine
- 3 tablespoons capers, drained
- Salt
- ½ pound baguette
- ¼ cup loosely packed, coarsely chopped fresh basil leaves
- 2 garlic cloves, chopped
- 4 salted anchovy fillets, rinsed and chopped
- 3 8-inch zucchini
- 12 small, round tomatoes

Toast the pine nuts in a small skillet over medium heat, stirring, until they are lightly browned and fragrant, about 13 minutes. Set aside.

Heat the oven to 400 degrees. Combine the olive oil and the onion in a large skillet over medium heat and cook until the onion softens, about 5 minutes. Add the minced garlic and cook until fragrant, about 3 minutes. Add the crushed tomatoes, wine, capers and ½ teaspoon salt and simmer until the sauce thickens, about 20 minutes.

Meanwhile, trim the crust from the bread and cut the bread into cubes. Place in a blender with the basil and chopped garlic and grind to fine, pale-green bread crumbs. Pour the bread crumbs into a medium bowl and stir in the anchovies and the pine nuts. Set aside.

Cut each zucchini in half crosswise and use a melon baller or serrated spoon to carefully remove some of the flesh from the center to make what looks like a canoe. Leave about ¼ inch of flesh at the sides and ends and a little more at the bottom. Discard the scooped out flesh. Season the inside lightly with salt and steam over rapidly boiling salted water until almost tender, about 15 minutes.

Cut a slice from the top of each tomato and use the melon baller or spoon to gently remove most of the pulp. Discard the pulp. Season the inside lightly with salt.

Pour the tomato sauce into a lightly oiled 5-quart gratin dish. Spoon the bread crumb mixture into the zucchini and tomatoes, mounding it slightly on top. You will need 1 to 2 tablespoons for each zucchini and 2 to 3 teaspoons for each tomato. Do not press the crumbs, or they will become pasty when cooked. Arrange the zucchini and tomatoes in the gratin dish.

Bake until the vegetables soften and the crumbs are browned, 30 to 35 minutes. Serve immediately, or set aside and serve at room temperature.

Summer Squash Stew with Pasta and Lima Beans

This dish comes together in about as much time as it takes to read the recipe. The combination of lima beans, yellow squash and tomatoes gives it a juicy vibrancy. Make sure you're generous with the freshly ground pepper.

6 SERVINGS AS A FIRST COURSE,
3 SERVINGS AS A LIGHT MAIN DISH

½ pound short tubular pasta
3 slices prosciutto, chopped
2 tablespoons butter
½ pound shelled lima beans (fresh or thawed frozen)
2 pounds yellow squash, cubed
¼ pound cherry tomatoes, cut in half
  Salt and freshly ground pepper
¼ cup grated Parmigiano-Reggiano

Cook the pasta in plenty of rapidly boiling salted water until almost tender, about 10 minutes. Drain and set aside, reserving some of the cooking water.

Combine the prosciutto, butter, lima beans and 2 tablespoons of the reserved pasta cooking water in a cold skillet. Cook, covered, over medium heat, stirring occasionally, until the beans begin to soften, about 5 minutes.

Add the squash, replace the cover and cook, stirring occasionally, until the squash softens and begins to glaze, about 5 minutes more.

Add the tomatoes and cooked pasta to the skillet and cook, stirring, until the tomatoes begin to melt into a sauce, about 3 minutes. Season with salt and pepper to taste. (Salt lightly at first; the pasta cooking water is salted, and you might not need to add much.) Dust with most of the grated cheese and divide among six heated pasta bowls. Garnish with the remaining cheese and serve immediately.

Tomatoes

. . . .

Few pleasures are simpler than sitting on your back step savoring the taste of summer's first vine-ripe tomato. Yet if you look at it closely, few things are more complicated. Sure, a bite of a sweet, deeply flavored tomato is one of life's more elemental joys. It's a flavor we yearn for all through the cold months, an experience we remember from our childhoods. And yet in real life, that kind of juice-gushing taste explosion is something that happens all too rarely. Tomato flavor, or the lack thereof, is one of the biggest complaints in the produce world. And there is no easy fix.

In part this is because the taste of a tomato is a devilishly difficult thing to pin down. In the first place, as real tomato lovers know, all tomatoes don't taste the same. Instead, there is a whole range of flavors, from almost citrusy to nearly meaty. But even if you're just trying to describe that essential tomatoey flavor, the problem is incredibly complex.

Flavors and aromas are made up of chemicals, and although some fruits, such as the banana, can be identified by just one (3-methylbutyl acetate, if you're curious), scientists who study flavor chemistry have identified more than four hundred compounds that go into the taste of a ripe tomato. And more than thirty of those are regarded as essential — detectable in amounts down to one part per billion. (Even this tiny bit may not be sufficient when you're dealing with tomatoes. The human nose can detect some odors/flavors in parts per trillion.)

Furthermore, the science of flavor is much more complicated

than simply putting a bunch of elements together in a chemistry set. When scientists talk about the chemical compounds that make up flavor, they don't speak the same language we do. They disdain romantic descriptions such as "beefy" or "minty" or "flowery" and instead zero in on specific names: Z-3-hexenal, beta-damascenone, beta-ionone and 3-methylbutanal.

Those are hardly words to make a gourmet's heart flutter, but they are the four flavor compounds found in the highest concentrations in a raw tomato. Z-3-hexenal is a fresh green fragrance often found in the aroma of cut grass (it also dissipates quickly, disappearing when heated). Beta-damascenone is generally described as fruity and sweet, while beta-ionone is fruity and woody. The last compound, 3-methylbutanal, is hard to describe; one source describes it as having "nutty-cocoa facets." It is also found in Parmesan cheese.

But a high concentration of a compound does not necessarily mean that it is the most important element. Some chemicals found in tomato flavor are more distinctive than others, even though they may be found at far lower concentrations. Methyl salicylate, for example, makes the distinctive smell of wintergreen. Linalool is a sweet, floral scent found in coriander seed and oranges. Safrole — found in wintergreen, anise, vanilla and camphor — is often described as smelling like root beer. In fact, it was used as a chemical flavoring for the drink until it was found to be mildly carcinogenic.

Chemicals are perceived in different ways as well. There are "top note" flavors, which tend to be heat volatile, meaning they are released just from the warmth of your mouth and are immediately identifiable. And there are "bottom note" flavors, which are not so volatile and linger in the background. And you wondered why there is no such thing as a good artificial tomato flavoring?

All this chemical name-dropping is of more than academic interest. To get a handle on the problem of tomato flavor, scientists need to be able to quantify just what it is made of, where it comes from and how it can be measured. This is far more complicated than you might think. To begin with, the chemical nature of flavor is not static. The flavor compounds that are present in a whole tomato change when the cell structure of that tomato is altered.

Slice a tomato, and chemicals combine and react. Chew it, and the same thing happens. Even more elusive: when you eat a tomato, some flavor compounds are formed by the interaction of the tomato's chemicals and your own — those enzymes found in saliva, for example.

The fruit is constantly changing, too. It doesn't take slicing or chewing to start the flavor reactions. Cell walls soften during ripening, which allows the exchange and combination of chemicals. This process, in particular, is the key to much of what we regard as fine tomato flavor. As tomatoes ripen, all sorts of flavor-building events occur. The first and most obvious is that the color changes from dark green to brilliant red (well, in most cases). This is not simply a matter of decoration. That color change is caused by the development of coloring chemicals called carotenoids. And unlike some other pigments, these have flavor-causing chemicals attached to them.

At the same time, as the tomato ripens, the amount of sugars builds, and their composition changes. Green tomatoes don't have much sugar, and what is there is mostly glucose. As the fruit ripens, the sugars that develop are much more heavily weighted to fructose, which is three times sweeter than glucose. The acids in the tomato change as well, from malic (such as that found in apples) to citric, which actually increases our perception of the sweetness of the glucose that remains. A whole host of aromatic chemical compounds are formed — hexanal (which is frequently described as "winey"), alanine and leucine (meaty) and valine (fruity). Furaneol, which has a characteristic pineapple aroma, is also created, as are some unusual aroma compounds that are not found in any other fresh fruits, but which are found in flowers. In addition, the sheer physical softening of the flesh that comes with ripening increases our perception of flavor.

This ripening doesn't have to happen entirely on the vine. Tomatoes are climacteric fruits, like peaches and mangoes, and if grown to something approaching physiological maturity, they can be picked while still solid green and continue to ripen in storage. And as with other climacteric fruits, this process is encouraged by exposing the fruit to ethylene gas. In the trade, most of the tomatoes ripened this way are sold as "mature-green," and they still make up the bulk of the standard tomatoes found in the market. Picked at just the right moment and ripened this way, a tomato can be adequate, but it will never be great. Picked a day or two too early, no matter how much gas is applied, it will never develop flavor, even though it may change color. This is particularly troublesome because it is nearly impossible to look at a whole green tomato in the field and tell whether it is mature enough to pick — the color is virtually identical to one that is still immature. The only way to tell for sure whether a green tomato is ready to pick is to cut it open and see whether the gel around the seeds has softened. Since the growers cut open only a few tomatoes to check and since tomatoes don't mature uniformly, inevitably a significant percentage will be so immature that they will never develop any ripe flavors at all. Even so-called vine-ripe tomatoes are picked at a stage that most of us would consider green, when only the very first traces of a tan, yellow or pink "blush" appear on the tomato's base.

The holy grails of tomato flavor study are techniques that would tell plant breeders which varieties have the best flavor, tell growers when tomatoes are ready to be picked and tell packers and shippers when the fruit is ripe enough to go to the store. Obviously, every tomato cannot be run through a sensory panel to determine its culinary worth. Just as obviously, taking apart a tomato's chemical components to try to measure flavor is incredibly difficult. One of the latest attempts to understand and measure just what is going on when we eat a ripe tomato was developed by scientists in Florida, who actually waved chemical sensors under the noses of people who were chewing tomatoes. Fine-tuning tomato flavor at this level is something that could take forever.

Much more practical steps can be taken that would improve tomato taste a lot more quickly, but the research to discover them is just beginning. Most of the research up until now has been aimed at solving what seemed to growers and shippers to be more important flaws.

At its most basic, the biggest challenge with tomatoes is the same one faced by so much of agriculture — to produce a fragile, temperamental product on an industrial scale. By nature tomatoes are thin-skinned and late-maturing, to say nothing of being prone to a wide array of diseases and insects. That we have such an insatiable demand for them — cut up in salads, sliced onto hamburgers, chopped up in salsas — is a cruel irony. It would be a lot better for everyone involved if those culinary functions could be filled by, say, an orange or a zucchini, any fruit or vegetable that isn't always poised on the edge of self-destruction.

But we must make do with the fruits and vegetables that we have, not the ones we may wish we had. And up to now, quite frankly, taste has been the least of anyone's concerns. As a result, breeders have developed tomato strains that resist all manner of cankers and wilts, without paying much attention to choosing ones with great flavor. And farmers and shippers still insist on picking fruit at the very earliest signs of impending ripeness to ensure that the tomatoes survive the trip to the grocery store. There hasn't even been any really reliable research that would clear up the most troublesome basic questions regarding tomato flavor: How much does it improve for every extra day the fruit is allowed to hang on the vine? How does this improvement work? Are there steps growers can take — such as using specific fertilizers or withholding water before harvest — that will make a difference in flavor?

One vital piece of the tomato quality puzzle involves steps you can take yourself. Cold temperatures wreak havoc with tomato flavor. The exact mechanism of how this works is still being studied, but without question it does happen. Temperatures below 60 degrees reduce the aroma-creating volatiles in the fruit. This change is quick and irreversible. If a case of tomatoes is even stacked next to a cooler for a day, that can be enough to ruin them. Tomato growers have been working diligently for the last several years to get that word out up and down the supply chain, but it is still not uncommon to walk into a grocery store and see tomatoes stored in the refrigerator. If this should happen to you, there is only one solution: turn right around and walk back out. That may not be much, but it is as close to a sure and simple fix as exists in the tomato world.

WHERE THEY'RE GROWN: Every year tomatoes are in a tight race with head lettuce as the biggest vegetable crop grown in the United States in terms of sheer dollar value. But whereas head lettuce is largely confined to one state (and, to an amazing extent, one region of one state), the tomato harvest is much more widespread: eighteen states grow commercially important amounts. Still, just two states — California and Florida — account for almost two thirds of the harvest. Florida dominates the winter market, while California rules in the summer. Tomatoes are one of the most heavily imported produce items. Roughly one third of our fresh tomatoes come from overseas, and Mexico produces as many tomatoes for the American market as either California or Florida. Oddly, Canada and the Netherlands are in second and third places, almost entirely with hothouse varieties.

HOW TO CHOOSE: Picking a terrific tomato is as much of an art as it is a science. Obviously, you should avoid any tomatoes with obvious flaws — dents, nicks or cuts. Also, you want tomatoes that are heavy for their size. You don't want a tomato that shows a lot of green, but beyond that don't pay so much attention to color, unless you're looking for a tomato to eat right away. Actually, overripeness can be as big a problem as underripeness. Overripe tomatoes are mealy and have off flavors, and you can tell them by their slack skin. In the end, trust your nose. Tomato aroma is the best indicator of quality.

HOW TO STORE: Do not, ever, put a tomato in the refrigerator. That will kill the flavor faster than anything. Store tomatoes in a cool, dry place away from direct light. If they are slightly underripe, they'll soften in a day or two.

HOW TO PREPARE: Tomatoes need to be peeled for cooked dishes. To do this, cut a shallow X in the blossom end and blanch it in boiling water until the skin begins to peel away. The time will vary according to the ripeness of the fruit — anywhere from 5 to 20 seconds. To seed tomatoes, cut them in half crosswise and squeeze out the seeds and pulp. Purists do this over a sieve, collecting the flavorful pulp in a bowl below.

ONE SIMPLE DISH: Dice seeded, unpeeled tomatoes as finely as you can. Dress them with a little olive oil, salt and freshly ground pepper. Taste them, and if they need a little red wine vinegar, add that. Place a log of fresh goat cheese on a plate and spoon the tomatoes around it. You may never go back to tomatoes and fresh mozzarella.

Heirloom Tomato Tart

Let's see — caramelized onions, buttery puff pastry, salty olives and cheese and great fresh tomatoes. What else could you want in a summer appetizer?

8 SERVINGS

- 1 17.3-ounce package frozen puff pastry
- 2 tablespoons olive oil, plus more for drizzling
- 2 onions, sliced moderately thin
- 2½ pounds tomatoes (preferably of different colors)
  Salt
- 1 large egg
- 2 ounces pitted oil-cured black olives
- 1 ounce pecorino Romano
  Torn fresh basil leaves

Remove the puff pastry from the freezer and defrost in the refrigerator for about 2 hours.

Combine the olive oil and onions in a large skillet over medium heat. Cook, covered, until the onions soften, about 15 minutes. Stir to make sure the onions aren't scorching, replace the cover and reduce the heat to medium-low. Continue cooking, stirring occasionally, until the onions are golden and sweet, about another 45 minutes. Set aside to cool.

Cut the tomatoes in half vertically, then slice each half horizontally as thinly as you can. Arrange the sliced tomatoes on a jelly-roll pan and sprinkle liberally with salt. Set the pan at a slant and let the tomatoes give up their liquid for at least an hour.

Heat the oven to 400 degrees. Remove the puff pastry from the refrigerator and unfold the sheets. Beat the egg until it is pale. Using a pastry brush, paint a thin strip of egg wash along 1 narrow edge of 1 pastry sheet. Arrange the second sheet so it is overlapping just along the painted edge and press to seal. Transfer the pastry to a baking sheet lined with parchment paper.

Scatter the cooled onions down the center of the pastry; they will not make a uniform layer. Drain the tomatoes. Arrange the tomatoes in overlapping slices in the center of the pastry on top of the onions. Drizzle a little olive oil on top of the tomatoes. Scatter the olives over the top, then shave the cheese over everything.

Fold the top and bottom edges of the pastry over to barely overlap the tomato filling. Fold over the sides in the same way. Cut a notch at each corner to reduce the thickness of the pastry there. Paint the edges with egg wash and bake until the pastry is puffed and dark brown, about 30 minutes.

Remove from the oven and scatter the torn basil leaves over the top. Let cool slightly before serving.

Golden Tomato Soup with Fennel

If you only know gazpacho as that liquid, finely chopped salad that's so often served, try following this basic technique — the real Spanish way — using any kind of good, ripe tomato. The soaked bread gives the soup a light creaminess that perfectly mellows the tomatoes' tart, rich flavor. If you don't have golden tomatoes, red will work.

Pimentón is a smoky Spanish paprika.

4 TO 6 SERVINGS

- 2 ounces crustless sturdy white bread, cubed
- 2 pounds golden tomatoes, chopped (see headnote)
- ½ cup chopped fennel (about ½ bulb)
- 2 tablespoons chopped red onion
- 1 garlic clove
  Fresh lemon juice (1 lemon)
- 1 tablespoon red wine vinegar
  Salt
- ¼ teaspoon ground cumin
  Dash Pimentón de la Vera or paprika
- 1 cup ice water
- ¼ cup plus 1 teaspoon olive oil
- ½ cup finely diced fennel (about ½ bulb)
- 3 teaspoons minced fennel fronds

Put the bread in a bowl and add enough water to cover. Let stand for at least 30 minutes to soften.

Squeeze the bread dry and put it in a blender with the tomatoes, chopped fennel, onion and garlic. Puree until smooth. Add the juice of ½ lemon, the vinegar, 1 teaspoon salt, the cumin and Pimentón and puree again. With the motor running, gradually add the ice water and ¼ cup olive oil. Chill well.

In a small bowl, stir together the finely diced fennel and the minced fronds. Moisten with the remaining 1 teaspoon olive oil and season with a couple of drops of lemon juice and a sprinkling of salt.

Before serving, strain the puree through a fine-mesh strainer if you want a perfectly smooth soup. Otherwise, whisk it to gently reincorporate anything that might have separated. Taste and add more salt, if desired. Divide evenly among four to six chilled soup bowls and garnish with 1 to 2 tablespoons of the diced fennel mixture. Serve.

Seared Scallops with Tomato Butter

This tomato butter is so insanely good that you could spoon it over wet cardboard and be happy. But when it is paired with sweet, crusty seared scallops, the combination is truly wonderful.

6 SERVINGS

- ½ **pound red and yellow cherry or grape tomatoes**
- 1¼ **teaspoons minced fresh tarragon**
- 4 **tablespoons (½ stick) butter**
- 1½ **teaspoons minced shallots**
- 2 **pounds sea scallops**
- **Salt**
- **Red wine vinegar (optional)**
- 2 **tablespoons olive oil**

Quarter each tomato lengthwise and combine in a bowl with the tarragon. Melt the butter in a small saucepan over medium heat. Add the shallots and cook until they soften and become fragrant, 2 to 3 minutes. Remove the butter from the heat and let cool for 2 to 3 minutes.

Prepare the scallops by removing the small, tough "foot" muscle that is attached to one side. Rinse the scallops and pat dry with paper towels.

When the butter is cool (it should be hot enough to soften the tomatoes slightly, but not so hot as to cook them through), pour it over the tomatoes and stir gently to combine. Be careful not to smash the tomatoes. Season with salt to taste (about 1 teaspoon). If the tomatoes lack acidity, add a few drops of red wine vinegar. Set the sauce aside while you cook the scallops.

Heat a large nonstick pan over high heat. Add the oil and heat until it is nearly smoking. Place the scallops in the pan and cook until they are browned and crusty on one side, about 3 minutes. Turn the scallops over and brown on the other side, about another 3 minutes. They should still be slightly translucent in the center.

Divide the scallops evenly among six plates and spoon a generous portion of the tomato butter over the top. Serve.

Cherries

. . . .

The roadside of modern agriculture is littered with the wreckage of once fabulous fruits that have been corrupted beyond recognition by the quest for commercial advantage. A rainbow of varieties that were treasured each in their own time and their own way has now been narrowed to only one or two choices, selected primarily not because of their eating quality, but because they happen to ripen at a specific time or ship better than the others. Judging strictly by the numbers, you could say that is exactly what has happened with cherries.

In the late 1900s, fruit scholar U. P. Hedrick filled an entire beautiful book with the hundreds of varieties grown just in the state of New York (more than two hundred kinds of sour cherries alone). Today the sweet cherry industry is dominated by only one variety. And as for finding fresh sour cherries, you might as well forget it. Even though more than 240 million pounds of sour cherries are harvested in an average year (almost all of them in Michigan), less than 1 million pounds are sold fresh; the rest are canned.

But a funny thing happened on the way to the cherry apocalypse. The one variety that ended up being chosen above all others is actually pretty darned good. Bing cherries from Oregon, Washington and California make up something like 65 percent of the fresh sweet cherry market (no other variety accounts for more than 10 percent). And at its best, the Bing is about as good as any cherry variety that has ever been grown — crisp on the outside, with a melting center that saves it from being crunchy; dark and sweet, with a nice tart backbone. Not only is the Bing delicious, it is also a legitimate antique. The variety was found in 1875 on the farm of eastern Oregon agricultural pioneer Seth Lewelling — as legend would have it, by a Chinese workman named Ah Bing.

In general, cherry varieties are lumped into four large categories: Bigarreaux (or Black cherries), Dukes, Hearts and Sours. Hearts and Dukes are exceedingly soft, and because they can't be shipped, they have largely disappeared commercially. Bigarreaux, or Blacks, dominate the fresh market. This family includes not only the dark red Bing but also the classic Royal Ann and its modern imitator Rainier, which blush from gold to crimson. (Royal Ann and Rainier cherries are the ones that traditionally have been dyed and preserved for maraschino cherries.)

Just as cross-country shipping changed the face of the cherry business in the twentieth century, international shipping may be doing it again today. The real gold rush in cherry farming is in getting the first fruit of the season to the Asian market. Jet-freighted to Japan and Hong Kong, fresh cherries fetch prices as much as ten times what they will bring at home two weeks later. As a result, growers are striving for earlier and earlier harvests. The typical Northwest Bing harvest begins in mid-June, but the real money is in getting those cherries to market as early as mid-May. Generally speaking, this means growing in warmer areas such as California's San Joaquin Valley, where the fruit ripen earlier.

This move south can be highly problematic, however. First of all, most of that area is at the extreme edge of the cherry's climatic range. Most cherry trees need at least seven hundred hours of chill every winter to go dormant. (That's the number of hours when the temperature is at or below 45 degrees.) Without that winter nap, the trees don't have the energy to produce great fruit. Higher elevations in the San Joaquin Valley can get that kind of chill usually but not always. Furthermore, cherries are extremely susceptible to rain — they crack and split — and winter in California can be very wet. Not only that, but cherries that ripen in too-hot weather are prone to problems such as spurring (the development of a little "beak" at the base of the fruit) and doubling (two cherries grown on one stem). So, if everything goes just right, California growers reap a windfall, but how often does everything go just right?

To a great extent, at least the last two problems have been

solved by the introduction of a new cherry variety called Brooks, a cross between the Rainier and a fairly nondescript but early variety called Burlat. Released by the University of California in 1988, the Brooks is no Bing, but it can still be a pretty good — though certainly not great — cherry. More important to anxious farmers, it resists the warm-weather flaws of earlier varieties and can be picked an average of ten days to two weeks before the Bing.

This has resulted in a massive shift in cherry growing. In the 1980s California ranked third behind Washington and Oregon. Between 1992 and 2002 California's acreage more than doubled, and in 2002 it climbed into first place (though production-wise it still lags behind Washington). California now accounts for more than a third of the fresh cherries grown in the United States (up from a quarter in 1992). At the same time, acreage in Washington has increased by more than half, but there growers are aiming at stretching the season longer into the summer, growing varieties such as Lapin, which now makes up about 10 percent of the national total. And in the big picture, total cherry acreage in the United States has increased from 45,000 in 1992 to 74,000 in 2003, while sales of cherry trees from commercial orchards almost doubled between 1999 and 2003. All of this leads some farmers to fear a potential glut in the market in coming years. Is it possible that eventually they will grow more cherries than we can eat?

The answer to that question is probably no, but wouldn't it be nice for us to see them try? Cherries mark the transition from spring to summer the same way strawberries celebrate the beginning of spring and pears the start of fall. Cherries are the first of the year's stone fruits to be harvested, and in many ways the sweetest, even if only because of that. Still to come are apricots, plums, peaches and nectarines, and even later almonds (yup, they're all cousins from the Prunus family). Like the rest of the clan, cherries rely for their flavor on the delicate balance between sweetness and acidity. That elusive thing called "cherry flavor" counts, of course, but it is hard to quantify. Most of the distinctive taste notes of cherries come from either almondlike benzaldehyde or clovelike eugenol. In taste tests, it seems to follow (but not necessarily be caused by) the accumulation of sugars: the sweeter the cherry, generally, the greater the flavor intensity.

How do you pick a sweet cherry? The key is color. Remember that most cherries on the market spring from the family called "Blacks." This is no accident. The darker red these cherries are, the more likely they are to be sweet (allowing for variation from season to season and orchard to orchard). The term of art that cherry growers use for a dead-ripe Bing is "mahogany," and that's a good description. With blushing cherries, such as Rainiers, the red will never get that dark, but the golden cheek should be deeply colored and not at all strawlike. Texture is another important aspect of the cherry's appeal. Unless you somehow happen upon a stray Duke or Heart (some are still out there, but mainly in antique gardens), cherries should be firm enough to be slightly crisp. Cherries can get overripe. When that happens, the acidity drops, resulting in a piece of fruit that tastes simply sweet. More important, the cherry softens at the same time. Overripe cherries tend toward a kind of flaccidness, as well as pitting and shrinking. The color is also more matte, rather than being bright and shining.

The types of cherries we get today have generally been favored for eating out of hand, and there is certainly nothing wrong with that. But they do cook well, too, though perhaps they lack the depth and tang of a great Montmorency (a sour cherry). This can be worked around: adding a dash of balsamic vinegar to cooked sweet cherries will add much of the "bottom" that might be missing. Don't overdo it — think of it as a subtle bit of makeup, not a mask.

When you're adding cherries to savory dishes — something that should be done only rarely and with due deliberation — use a good red wine vinegar instead. Happily, heating cherries intensifies their perfume, and even more so if the pits are left intact. (The pits of stone fruits are rich in almond scent. In fact, they can be used as substitutes for hard-to-find bitter almonds. Of course, if that's your aim, you're better off using fruits such as peaches or apricots, which have bigger pits.) In many traditional recipes, the stones are left in cherries that are cooked. Honestly, though, the improvement in taste is pretty minimal. The real motive probably has more to do with limiting the amount of bright red juice the cherries will bleed into the dish. Basically, it comes down to weighing the advantages of authenticity against the possibility of a chipped tooth for you or someone you happen to be feeding.

WHERE THEY'RE GROWN: Most cherries are grown in California, Oregon and Washington. In California they're grown primarily in the higher elevations of the Central Valley. In Oregon they're grown in the Willamette Valley and the Columbia River basin. In Washington they're grown mainly on the dry eastern plains.

HOW TO CHOOSE: The most common cherry is the Bing, and in the cherry box at the store, there will be a wide range of ripeness. Take the time to choose carefully, sorting through a small handful at a time. Choose fruit that is dark red, almost to the point of being black, and shiny, not matte. There should be no shriveling or wilting. Common faults such as doubling and spurring do not affect the flavor.

HOW TO STORE: Stored tightly wrapped in a plastic bag in the coldest part of your refrigerator, cherries will last a surprisingly long time — up to 2 to 3 weeks. Do not wash until just before using.

HOW TO PREPARE: Cherries should be stemmed and usually pitted.

ONE SIMPLE DISH: Macerate 1 pound pitted cherries and ⅓ cup sugar in a bowl until the cherries release their juice, 30 to 45 minutes. Add ¼ cup dry red wine and set aside until the cherries absorb it, another 15 minutes. Tie together in a cheesecloth bundle 1 clove, 1 whole allspice and a 1½-inch cinnamon stick. Cook the cherries, the juice and the spice packet in a wide nonstick skillet over high heat until the juice begins to thicken, about 10 minutes. Remove from the heat and stir in 1½ teaspoons balsamic vinegar. Let cool and serve over vanilla ice cream.

Cold Spiced Cherry Soup

An adaptation of a classic Hungarian dish, this slightly sweet soup makes an elegant first course for an early-summer dinner.

6 TO 8 SERVINGS

- 12 whole allspice
- 12 black peppercorns
- 8 whole cloves
- 1 3-inch cinnamon stick
- 1 750 ml bottle rosé
- 2 cups water
- ⅓ cup sugar
- 2 pounds cherries, stemmed and pitted
- 1 tablespoon balsamic vinegar, plus more to taste
- ¼ teaspoon almond extract
- ¾ cup crème fraîche or yogurt, plus more for serving

Cut a piece of cheesecloth about 5 inches square. Place the allspice, peppercorns, cloves and cinnamon in the center. Bring the corners together and tie the packet closed with a piece of string. (Alternatively, you can place the spices in a large tea ball.)

Bring the wine, water and sugar to a simmer in a large saucepan over medium heat and add the spice packet. Cook until the wine loses its raw alcohol smell and mellows in flavor, about 15 minutes. Cooking gently will help the wine retain its delicate fruitiness.

Add the cherries and simmer gently until slightly soft, about 15 minutes. Remove the pan from the heat and stir in the balsamic vinegar and almond extract. Taste and adjust the vinegar; the soup should be slightly sweet, but with a definite tart backbone. Transfer to a lidded bowl and refrigerate until well chilled, at least 4 hours.

Just before serving, discard the spice packet and add the ¾ cup crème fraîche. Replace the lid and shake gently to mix well. The soup should be lightly creamy and pale pink.

Ladle the soup into chilled bowls and garnish each bowl with a small dollop of crème fraîche swirled into the center. Serve.

Red Wine–Poached Cherries

This unusual technique for poaching cherries comes from chef Josiah Citrin of Melisse restaurant in Santa Monica, California. Cooked in resealable plastic bags, they stay firmer than they normally would. And because there is less cooking liquid, the fruit itself seems to come to the fore, so the cherries taste more like cherries, as does the wine syrup that's left over.

Serve over vanilla ice cream.

4 SERVINGS

1 **cup dry red wine**
2 **tablespoons sugar**
1 **pound cherries, stemmed and pitted**

Gently simmer the red wine and sugar in a small saucepan until the wine loses its raw alcohol smell, about 10 minutes. Set aside to cool to less than 140 degrees.

Place the cherries in a 1-quart resealable plastic bag and pour the cooled wine over them. Press out all the air and seal the bag tightly. Place this bag inside another, press out the air and seal tightly again.

Bring a large pot of water to 140 degrees. Place the double-bagged cherries in the water and poach at between 140 and 150 degrees for 20 minutes. During the first 5 to 10 minutes, you'll need to pay attention to the temperature, but after that it'll maintain without much fussing. (If the water gets too hot, just add a little tap water to bring the temperature down.)

After 20 minutes, remove the double bag from the pot. Place a strainer over a small saucepan and empty the cherries into it, collecting the poaching liquid underneath. Set the cherries aside in a bowl and simmer the poaching liquid until it has reduced to a syrup. Pour the syrup over the cherries and toss to coat well before serving.

Cherry-Almond Cobbler

The topping is cakey, almost like an almond torte, not the familiar crisp biscuit. Serve this with vanilla ice cream or a splash of lightly sweetened heavy cream or crème fraîche.

Almond meal is nothing more than finely ground almonds and can be found in many grocery stores. If you can't get it, you can grind the almonds yourself in a food processor. Grind them to the texture of coarse cornmeal or couscous, but do it in short pulses so you don't turn the nuts to butter.

4 SERVINGS

- 1 **pound cherries, stemmed and pitted**
- **Sugar**
- 1 **teaspoon grated orange zest**
- 1¼ **cups (4 ounces) ground almond meal (see headnote)**
- ¼ **teaspoon salt**
- 2 **tablespoons butter**
- 3 **large eggs, separated**
- 2 **tablespoons orange liqueur**

Heat the oven to 400 degrees. Put the cherries in a small (4-cup), well-buttered gratin dish. Sprinkle with 1 tablespoon sugar and the orange zest and stir to combine. Roast the cherries until they soften slightly and become fragrant, about 10 minutes.

In the meantime, grind together the almond meal, ⅓ cup sugar, the salt, butter, egg yolks and orange liqueur in a food processor.

Beat the egg whites until they hold soft peaks, then add 1 teaspoon sugar and beat until they hold stiff peaks.

Add one third of the egg whites to the food processor and pulse to incorporate into the almond meal mixture. Spoon the mixture into one side of the bowl with the egg whites, spread the egg whites over the top and fold them into the mixture until light and somewhat flowing. There may be small patches of unincorporated egg whites; it's better to leave them than risk overfolding the batter and deflating it.

Spoon the mixture over the roasted cherries and sprinkle with 1 teaspoon sugar. Place the gratin dish on a baking sheet to catch any spills and return to the oven. Bake until the top is puffed and well browned, 15 to 20 minutes. Serve warm or at room temperature.

Grapes

· · · · · · · ·

By almost every measure imaginable, grape growers have been amazingly successful. Although national consumption of fresh fruits and vegetables barely manages to stay even from year to year, American consumption of fresh grapes has more than tripled since 1970, to more than eight pounds per person. What's particularly astonishing is that the only fruits we eat more of — bananas, apples and oranges — are all easily shipped and able to withstand long storage. Grapes, however, are the very picture of fragility: nothing more than thin-skinned little bags of juice.

Another amazing aspect of the grape-growing business is just how concentrated it is. Those billion dollars worth of fresh grapes America consumes every year are grown almost entirely in two small areas of California. In the winter, grapes are grown in the Coachella Valley, a couple of hours east of Los Angeles. In the main summer season, they are grown in a narrow band of the San Joaquin Valley from just north of Bakersfield to just south of Fresno. Those two areas account for roughly 97 percent of the fresh grapes grown in the United States.

Sadly, the one place the industry has been consistently less than amazing is in growing a grape that has any flavor. Table grapes today have all the personality of teen idols. They are essentially nothing but guilt-free snack foods — conveniently packaged sugar water that allows you to feel virtuous while you eat it. Grapes have become wildly successful by eliminating anything that could turn consumers off — such as distinctive flavor.

Within the past couple of years, though, growers and breeders have been working on changing that. Ironically, they're concentrating on recapturing the flavors of one of the world's oldest and best loved grapes: the Muscat.

Muscat grapes haven't always been as innocuous as they are now. In fact, hard as it may be to believe, not so long ago certain varieties were actually known for their flavor. Today you could sample half a dozen commercial grapes with your eyes closed and barely be able to tell one from the other, red from green from black. Slip an old-fashioned Concord into the mix, or a Muscat, and the picture would change entirely. These are grapes with character; there is no mistaking one of them for another.

The taste of the Concord, an American native, is familiar to most Americans — whether or not they have actually ever seen the grape. It's the flavor they know so well from grape juice, grape jelly and grape jam (and Mogen David wine). When grape growing was still centered in New York, this was the dominant variety. Concord grapes have a distinctive flavor that is usually described in wine circles as "foxy," whatever that means. For those with a more chemical bent, the smell of a Concord is that of the chemical ester methyl anthranilate.

It is odd that the grape that was the inspiration for "grape flavor" is now so difficult to find. At the turn of the last century, Concord was king. Much of it was eaten out of hand, fresh, but much of it also went to the burgeoning juice business, which got its start in New Jersey. There, in 1869, a dentist named Thomas Welch began experimenting with the newly developed process of pasteurization. He found that by using it, he could make a grape juice that wouldn't turn into wine.

Although the Muscat grape isn't as widely known as the Concord, it is more complex, full of allusion even on a chemical level. It comes from a mix of compounds called terpenoids: geraniol (also found in nutmeg and ginger), linalool (also found in flowering herbs such as lavender and jasmine), and nerol (also found in orange blossoms and cardamom). This heavenly floral perfume is probably most familiar from wines that are made from the grape: the Italian Moscato and Moscato d'Asti and the French Muscat de Beaumes de Venise. Muscat grapes are seldom used in American wines, although maverick wine producer Randall Grahm makes a delicious dessert wine from it called Vin de Glacier.

You might wonder about eating other wine grapes out of hand — Cabernet Sauvignon, Chardonnay, Pinot Noir. Actually, these have never been very popular fresh. Winemakers value different attributes than grape eaters, including thicker skins (for better color) and a lower ratio of sugar to acid (less sweet and more tart). Although these qualities can make exquisite wine, they do not appeal to the popular palate.

A true grape fan can also mourn the loss of other great old-time grapes such as Chasselas Doré, Italia, Lady Finger and Ribier. Almost all of them have now vanished — at least as far as commercial table grape production is concerned. The dominant table grapes today are varieties such as Thompson Seedless, Ruby Seedless, Crimson Seedless and Flame Seedless. Their primary selling point is self-explanatory. It's hard to be a popular convenience food when people have to interrupt their snacking to spit out the pips. This is not to say that all seedless grapes are by nature bad. Thompson Seedless (the familiar green grape) is a prime example.

Originally called the Sultanina Bianca, it was popularized in this country by William Thompson, a California nurseryman in the late nineteenth century. By far the most widely planted table grape in the United States (this one variety accounts for more than 25 percent of the total acreage), the Thompson Seedless is also the dominant raisin grape and is used (though almost never credited) in making inexpensive wines.

The fact that it has become something of a poster child for flavorless grapes is due much more to poor handling than to poor genetics. Properly grown and matured to full ripeness, the Thompson Seedless has startlingly good flavor. Perhaps it does not have quite as much character as the Concord or Muscat, but it does have a pleasingly flowery quality. The problem is that you can rarely find a fully mature Thompson Seedless in the market. When ripe, the variety has a tendency to "shatter" — that is, the grapes fall off the bunch. This is inconvenient for the grower, the retailer and the consumer. And so Thompsons are usually picked when they are still green. At this point they can be sweet, but they are never much more. Still, if you ever happen across a bunch of Thompsons that are amber-gold in color, snatch them up and see what you've been missing.

The modern absence of high-flavored grapes is not due to a lack of effort by plant breeders. Every couple of years, it seems, another great green hope is unveiled — a grape with distinctive flavor that will meet all the commercial requirements of growers and retailers. Almost inevitably, this new hope is found wanting in some way. One of the most recent attempts is a grape called Princess, introduced in 1999. This is a large green grape that, when decently ripened, has a haunting Muscat flavor. Like an Internet IPO, it was jumped on early by farmers, whose ardor then abruptly cooled. It turns out that just as those first vines were coming into full bearing, problems emerged. The grape is extremely unreliable in terms of production — depending on the early weather, farmers could have a great year or a lousy one. And so they've moved on to the next hopeful variety.

In a big-money industry like grape growing, any possible advantage must be exploited. This goes for even the most popular grapes. The Thompson Seedless, for example, is coddled like some kind of exotic bonsai tree. Not only are the vines trained to grow along specially designed trellises, but they are also meticulously pruned to manage the right number of leaves, the right number of shoots and the right number of grape clusters. And that's just the start.

Left to its own devices, Thompson Seedless vines produce grapes that are quite small, particularly when picked early. To get around that, farmers have come up with some innovative techniques to increase grape size. The first is called "girdling," and it involves cutting a ring in the bark all the way around the base of the vine just as the grapes begin to emerge (and often again as the grapes begin to gain color). This interrupts the flow of nutrients to the leaves and concentrates them in the fruit. Another technique is the application of gibberellic acid ("gibbing"). This is a naturally occurring plant growth regulator extracted from a cultivated fungus. (Some forms of it qualify for organic use.) It increases the size of individual grapes and also stretches out grape clusters, allowing for better air circulation and thus reducing disease. Between girdling and gibbing, a farmer can increase the size of an individual grape by as much as a third.

Certainly, not every aspect of the grape business is so high-tech. In fact, in many places raisin production hasn't changed since it was first introduced in the late nineteenth century. It might surprise you to know that until the 1870s, raisins — about as commonplace a food as there is today — were regarded as pricey exotics (they had to be imported from the Mediterranean). But the same happy combination of heat and lack of humidity that made growing grapes so easy in California's Central Valley also proved perfect for drying them. Today, as then, most raisins are made by the sun. Long strips of white paper are laid down between the rows of grape trellises. When the bunches are picked, they are placed on the paper and left to dry for two to three weeks. Almost all raisins are made from Thompson Seedless grapes. The difference between light (golden) and dark raisins is sulfuring. Golden raisins are sprayed with a sulfur dioxide compound before drying to keep their color from changing.

There are several varieties of specialty raisins as well. You can find varietally labeled raisins at farmers' markets and gourmet stores — mainly Muscat, Flame, Crimson and Sultana. There is also a tiny raisin called a currant, which is the cause of no end of confusion. It is commonly believed that these are derived from the small berries that are called currants (mostly either bright red or black fruit that are grown almost entirely in the northern British Isles). But those currant fruits have nothing to do with currant raisins, which are made from the tiny Black Corinth grape, and if you say Corinth with the accent of a New York produce dealer, you will understand the root of the confusion. These are sometimes called Zante currants, which alludes to the Greek island from which the grapes were first imported. California-grown Black Corinth grapes are also becoming popular fresh. They are usually labeled Champagne grapes, but of course they are not used for champagne at all. Their tiny size is very cute, but they don't have much flavor.

WHERE THEY'RE GROWN: Almost all of the commercial fresh table grape production in the United States takes place in California: the southern end of the San Joaquin Valley and the desert winter growing fields near the California-Arizona border.

HOW TO CHOOSE: Grapes should be heavy for their size, with taut, firm skins.

HOW TO STORE: Refrigerate in a tightly sealed plastic bag. Don't wash grapes until just before you use them, as surface moisture will break down their thin skins. If the grapes are moist when you buy them, slip a paper towel into the bag.

HOW TO PREPARE: Give grapes a quick rinse in ice-cold water just before serving and immediately pat them dry.

ONE SIMPLE DISH: Pair a bunch of grapes with a selection of fresh cow's and goat's milk cheeses for a late-summer dessert.

Pan-Crisped Duck Breast with Roasted Grapes

Even mild-flavored grapes gain character when you roast them briefly in a hot oven. And that intensification is just right for the richness of a perfectly cooked duck breast.

4 SERVINGS

- 4 **duck breasts (⅓–½ pound each)**
  Salt
- 1 **teaspoon black peppercorns**
- 2 **whole cloves**
- 2 **whole allspice**
- 4 **small bunches seedless grapes**
  Vegetable oil
- 1 **bunch fresh thyme**

With a sharp knife, cut a shallow cross-hatching on the skin side of the duck breasts, through the skin but not through the fat to the meat. In a spice grinder or with a mortar and pestle, grind 1 tablespoon salt with the peppercorns, cloves and allspice to a fine powder. Season the breasts liberally on both sides with the mixture, place them on a plate, cover them tightly with plastic wrap and refrigerate until ready to cook.

Heat the oven to 400 degrees. Place the grape bunches in a bowl and add just enough oil to coat them lightly, 2 to 3 tablespoons. Sprinkle with a little salt. Toss gently to coat lightly. Place the grapes in a roasting pan and scatter the thyme over the top. Roast until the grapes are fragrant and just about to split open, about 10 minutes. Remove from the pan and keep warm until ready to serve.

When ready to cook the duck, heat 1 tablespoon vegetable oil in a non-stick pan. When it is almost smoking, pat the skin side of the duck breasts dry with a paper towel and place the breasts skin side down in the hot pan. Sear until the skin is a deep golden brown, 7 to 10 minutes. Turn the breasts and cook on the other side until they are medium-rare in the center (135 degrees on an instant-read thermometer), 3 to 5 minutes more.

Transfer the duck breasts to a carving board and cut on the bias into thick crosswise slices. Arrange a sliced duck breast on each plate, garnish with a bunch of roasted grapes and serve.

Melons

· · · · · · · ·

Choosing the right melon is one of the more confusing
rites of summer — and you probably don't know the half of it. Some
people say you should thump melons. Some say you should give
them a sniff. Some claim the secret is all in the skin. Some tell
you to play with their bellybuttons (the melons', not the people's).
They're all right, and they're all wrong. It all depends on what kind
of melon you're talking about (and, come to think about it, just exactly what you mean by "melon").

It will come as no surprise to anyone who has paid melons more
than a passing glance that they are members of the gourd family,
along with squash and cucumbers. Collectively, these are known
as cucurbits. The specific genus that includes melons (well, most
melons) is *Cucumis*. As you can probably tell by the name, it also
includes cucumbers.

Within the *Cucumis* genus, melons are subdivided into several groups — just how many depends on whom you're talking to.
The first group is called Inodorus (the name literally means "without smell"), or winter melons, even though they are harvested in
the summer, just like the others. They include large melons such
as the casaba and honeydew. They usually have green flesh, but
not always (there are orange-fleshed honeydews, which are very
good, too). Inodorus melons have fairly crisp, slightly grainy flesh
and tend to be very sweet, with a slightly honeyed quality to the
flavor.

The second important group is Cantalupensis. These are smaller melons with scaly skin and usually orange, melting flesh. They tend to have a highly floral, slightly musky flavor. As you can probably guess, these include cantaloupes. But wait: what you probably think of as a cantaloupe isn't a cantaloupe at all; it's a muskmelon. A true cantaloupe is a melon such as the French Charentais or Cavaillon. ("Cantaloupe" is the "Frenchification" of the Italian Cantalupo, which was the name of the pope's summer estate outside Rome. Supposedly, one of those fifteenth-century "gourmet popes" had the seeds for these brought from the Near East.) Probably the most common true Cantalupensis melon in the United States is the Israeli import Ha-Ogen.

In fact, that melon you think of as a cantaloupe might not even be considered part of the Cantalupensis family — depending on which botanist you ask. Some scientists and growers recognize a third class of melons called Reticulatus. These differ from Cantalupensis melons because their skin is netted rather than scaly. Other than that, they're pretty much the same, although their aroma can be a little more floral — hence the name muskmelon.

It's not really important that you know the ins and outs of all these different families — at the rate that farmers are experimenting with new varieties, that's quite a challenge. What is important is that you recognize the difference between smooth-skinned and rough-skinned melons, because you select these melons in very different ways.

Rough-skinned melons are the easiest to choose, because they give you so many clues. The first thing to check is the netting or scaling. It should be tan or golden in color and definitely raised above the background skin, which should be golden in color, not green. Some rough-skinned melons are also ribbed. In a mature melon, those ribs are more pronounced.

Inspect the skin for the pale spot the French call the *couche,* which is the place where the melon rested on the ground. It should be creamy or golden and pronounced, but ideally not too much so. If there is no *couche,* the melon may have been picked too early. If the *couche* is too big, the melon rested in one place for too long. Really good farmers turn their melons so no single spot touches the ground for the entire time. A clean bellybutton is important, too. All rough-skinned melons are harvested at what farmers call "full slip," which means that the fruit slips cleanly away from the vine, leaving no trace of a stem in the bellybutton. Any stem at all indicates that the fruit was harvested too early. One of the best ways to choose a rough-skinned melon is also the most obvious: give it a sniff. When fully ripe, these melons develop a heavenly, musky floral perfume that you can smell at the other end of the produce section.

Sadly, the only one of these clues that works for smooth-skinned melons is the *couche*. These fruits are devilishly hard to choose. They don't have netting, so you can't check that. They don't "slip" from the stem, so the bellybutton is no help. And they usually don't have a smell. You don't have to be psychic to choose a good melon, but you do have to be extraordinarily sensitive. (Come to think of it, a little ESP couldn't hurt.)

The first thing to look for is color. This is extremely subtle, the difference between a "hard" green or white and a more golden "creamy" color. If you look at several melons, you'll see the distinction. When these melons are fully mature, they also develop a slightly waxy texture.

The best indicator of quality I've found in smooth-skinned melons is what growers call "sugar spots." These are brown flecks on the surface. Unfortunately, you'll see them only at farmers' markets. Supermarket produce managers tend to regard them as imperfections and wash them off.

So now you know how you choose a mature melon, but remember that there is a difference between ripeness and maturity. Melons continue to ripen after picking — the flesh softens, and the aromas and flavors become more intense — but they don't get any sweeter. This softening is usually most evident at the blossom end of the fruit. Press gently: if there is a little give, the melon is ripe; if you have a melon that still feels very firm, leave it at room temperature for a couple of days.

So much for the *Cucumis* melons. The elephant in the room that hasn't been discussed is the watermelon, perhaps the most popular melon of all — at least in the United States. Watermelons belong to a different branch of the Cucurbit family: *Citrullus*. Their closest relative is the bitter apple, a small, hard fruit that can be poisonous in moderate doses. It used to be that all watermelons were red, had seeds and were gargantuan in size. But there has been a virtual arms race in watermelon breeding over the past several years, and now we have watermelons that are yellow, watermelons that have no seeds (actually, they do, but they're few and underdeveloped) and watermelons that are built for two rather than two hundred. Recently, breeders have even come up with watermelons that are the perfect shape and size to fit in a refrigerator.

You select all of them the same way, which is pretty much the way you choose a smooth-skinned melon. Check the *couche,* which should be well developed, and check the skin, which should have a slightly waxy quality. If you have a good ear, you can try the thump test, too. This is the traditional way to choose a watermelon, but I think it's the hardest to get right. Rap the melon near the center with your knuckles. With a ripe melon, there will be a certain resonance: it will sound like knocking on a hollow-core door.

WHERE THEY'RE GROWN: Roughly two thirds of the spring harvest of cantaloupes is pretty evenly divided between Arizona and California, with Georgia and Texas chipping in the rest. Spring honeydews come mostly from California, with some from Texas, and spring watermelons come from Florida and Texas. In the summer California's Central Valley harvest kicks into high gear. Mendota, a small town on the valley's western slope, bills itself as the Cantaloupe Center of the World. California produces more than three quarters of the nation's summer cantaloupes and all but a smattering of its honeydews. Georgia and Texas are the dominant states for summer watermelons, with California and South Carolina also contributing important amounts.

HOW TO CHOOSE: Good cantaloupes are deeply scaled or netted and have a golden background color, clean separation from the stem and a deep floral fragrance. Good honeydews have a creamy color and a waxy texture to the skin. Good watermelons have a waxy skin and sound hollow when thumped lightly.

HOW TO STORE: Store all melons at room temperature. Cantaloupes and honeydews continue to ripen after being picked. This ripening makes them more fragrant and complex, but it doesn't make them any sweeter. If you prefer your melons chilled, put them in the refrigerator overnight. Much longer than that, and you'll see the rinds begin to pit and decay.

HOW TO PREPARE: Split cantaloupe and honeydew melons in half and spoon out the seeds. Watermelons need only be cut into sections.

ONE SIMPLE DISH: Melons are sweet, but they are surprisingly good when paired with salty or peppery companions. Wrap chunks of honeydew or cantaloupe in prosciutto or other thinly sliced cured meat.

WHAT TO REFRIGERATE

For most of us, putting food in the refrigerator is a reflex, not a considered act. We get home from the grocery store, sort out the packaged goods and shove everything else in the fridge — potatoes and peaches, berries and basil, all treated alike. But before you do this, stop and pay attention to what you're doing. In some cases, it takes only a day at room temperature to ruin a fruit or vegetable. In other cases, it takes only a couple of hours of cold. And in a few cases, the question of whether to refrigerate or not depends on the circumstances.

The best first step is understanding what a refrigerator does and why it can be important. Even after picking, fruits and vegetables continue to take in oxygen and give off carbon dioxide and heat. For most produce items, chilling slows the rate of that respiration. Generally, the closer you can come to 32 degrees, the more slowly respiration will occur. (That is the freezing temperature of water, but most fruits and vegetables won't freeze until several degrees colder because of the sugar they contain.) Different varieties of fruits and vegetables respire at different rates, ranging from those that hardly breathe at all (dates and nuts) to those that seem to be almost panting (asparagus, mushrooms, peas and corn).

At the same time, fruits and vegetables give off moisture, which is called "transpiration." Slowing transpiration is the purpose of the refrigerator's small, tightly sealed crisper. Looked at on a cellular level, most plant material is predominantly made up of water held in little cellulose sacks. When those sacks are full, the fruits and vegetables are firm. When the sacks start to lose moisture, the fruits and vegetables soften and wilt. Different fruits and vegetables transpire at different rates, roughly equivalent to their rates of respiration. Some have thick skins that slow the rate, such as apples, beets, hard-shelled squash, potatoes and citrus fruits. Some have hardly any peel at all, most notably lettuces and other greens.

203

If it were as simple as remembering the relative rates of respiration and transpiration, the whole question of whether to refrigerate would be a lot easier. But there is another layer of complexity. In some fruits and vegetables, chilling actually causes physiological damage. Refrigerating a tomato, for example, breaks down the chemical compounds that give the fruit its flavor and fragrance. Once chilled, a tomato may look just as pretty, but it will never regain its flavor. Potatoes convert starch to sugar if refrigerated and take on a sweet taste. Some fruits and vegetables that suffer chill damage might surprise you. Cucumbers, for example, develop soft spots on the surface. So do eggplants. Although most leafy vegetables and herbs need to be refrigerated, chilling wipes out basil, turning it black within a couple of hours.

Some fruits and vegetables can be refrigerated only in certain situations, such as after they're fully ripe. Peaches that have a dry, pithy, cottony texture and weak flavor have suffered chill damage as a result of being stored at the wrong temperature before they're fully ripe (see page 117).

It may be too obvious to mention, but any fruit or vegetable that has been cut up must be refrigerated, even if it is something you would normally leave out. Although chilling may damage it in some ways, it is far better than risking the spoilage that will come so quickly if it is left at room temperature.

This is a lot to remember, even for professionals. The tomato and stone fruit industries in particular have intensive programs to try to educate workers in warehouses and grocery stores about proper handling. Probably the easiest thing to do is to make a copy of the following list and keep it on your fridge (along with all those art projects and report cards). You should refrigerate all fruits and vegetables *not* listed here.

Never Refrigerate

BANANAS AND PLANTAINS

POTATOES AND SWEET POTATOES

STORAGE ONIONS AND GARLIC

TOMATOES

Refrigerate Only Briefly (no more than 3 days)

CUCUMBERS

EGGPLANTS

MELONS (only after fully ripened)

PEPPERS

Refrigerate Only After Fully Ripened

AVOCADOS

PEACHES, PLUMS AND NECTARINES

PEARS

Melon Salad with Shrimp and Arugula

This is an extension of the idea of pairing melons with peppery tastes. In this case, the spice comes from arugula. If you can get wild arugula from a farmers' market, use it: it has a more pronounced peppery flavor and a firmer texture than cultivated arugula. In a pinch you can use cooked shrimp and eliminate the cooking liquid from the dressing.

6 SERVINGS

- 1 cup water
- ½ cup dry white wine
- 2 tablespoons sherry vinegar
- Salt
- Pinch red pepper flakes
- 1½ teaspoons minced shallots
- 1 pound shell-on medium shrimp
- ½ 4- to 5-pound cantaloupe, seeded
- ¼ pound arugula (preferably wild; see headnote)
- 2 tablespoons fresh lemon juice, plus more to taste
- ⅔ cup olive oil

Bring the water, white wine, vinegar, 1½ teaspoons salt, the red pepper flakes and minced shallots to a simmer in a saucepan over high heat. When the liquid boils, add the shrimp. As soon as the liquid returns to the boil, cover it tightly and turn off the heat. When the pan is cool enough, put it in the refrigerator.

When you are ready to serve the salad, peel the shrimp, reserving the cooking liquid. Cut the cantaloupe into quarters, then cut the flesh away from the rind into ½-inch-thick slabs. Trim the arugula and tear into bite-size pieces.

Strain the shrimp cooking liquid. Place 2 tablespoons of it in a small, lidded jar and discard the rest. Add the lemon juice and olive oil and shake well to make a smooth, thick dressing. Taste for salt and lemon juice.

Put the melon slabs in a bowl and add enough dressing to coat lightly (1 to 2 tablespoons). Toss gently and arrange in a single layer on a serving platter. Add the arugula to the bowl and add enough dressing to coat lightly (1 to 2 tablespoons). Toss and arrange on top of the melon. Repeat with the shrimp and arrange on top of the arugula. Serve immediately.

Honeydew Ice Parfait with Blackberries and White Port

This parfait works well with Muscat-based wines such as Moscato or Beaumes de Venise. But when I tried it with a good-quality white port, such as is made by Ramos Pinto, I knew I'd found the perfect match.

6 SERVINGS

- 1 4- to 5-pound honeydew melon
- 3–4 tablespoons sugar
- 1 pint blackberries (or substitute blueberries or raspberries)
- 2 tablespoons white port

Cut the melon in half and scoop out the seeds. Using a spoon, scoop out chunks of melon flesh, putting them in a food processor. Don't dig too deep — the melon close to the peel has a strong cucumber flavor. You should get about 1½ pounds of melon chunks.

Puree the melon and stir in 2 tablespoons sugar. Taste and add another tablespoon if necessary. Too much sugar will mask the melon's flowery qualities, but because chilling reduces the flavor, the mixture should be very sweet.

Pour the puree into a large, shallow metal pan, such as a baking pan, that will hold it to a depth of ¾ to 1 inch. Freeze for 30 minutes. Remove the pan from the freezer and stir the puree with a fork, breaking up any chunks of ice. Repeat 4 or 5 times over 2 to 3 hours. Each time the ice will be a little less liquid and will stick together more. When it is firm enough to hold a shape, it is done.

Try not to let the melon ice freeze solid. If it does, chop it into small pieces in the pan and grind it in the food processor. The result will be lighter and fluffier, and the flavor will not be as dense and luscious.

Stir together the berries, 1 tablespoon sugar and the port. Spoon the ice into six glasses (martini glasses are perfect; short wineglasses are nearly as good) and spoon some of the berries with their juice over the top. Serve.

Peaches and Nectarines

. . . .

Just as a rose is not merely a rose, a peach is not merely a peach, and a nectarine certainly isn't merely a nectarine. Sure, that's what the sign says when you go into the grocery store. But that's for the convenience of the grocer, not a clue for you. The reality is that dozens, even hundreds, of different varieties of peaches and nectarines are grown and sold commercially every summer, sometimes of widely different qualities. Most of them are harvested for only a week or two, and then they're gone. To stitch together something approaching a continuous season-long marketing campaign for fruit, grocers sell many different varieties under the name "peach" or "nectarine." But that simple labeling can disguise what may be deep and fundamental differences among the different fruits.

The most obvious difference is the one between clingstone — fruit in which the flesh is glued to the seed inside — and freestone — that in which the seed floats free. For the most part, this difference is largely academic for grocery shoppers. Whereas clingstones once dominated the fresh peach market (and, in fact, are still grown in almost equal quantities as freestone), they are now almost entirely relegated to the canning industry. Customers just don't want to have to wrestle the flesh away from the seed anymore. It's messy.

And that's how it goes in stone fruit land. Plant breeders are trying all the time to come up with new twists on old favorites. Frequently, the changes are designed to meet the demands of the growers or packers. One very delicious family of peach varieties has nearly vanished commercially because it forms a small but definite "beak" at the bottom of the fruit. That little point tends to break during packing and shipping, opening the door to spoilage.

This is certainly not to say that the wants of consumers are ignored. In fact, they drive some of the most fundamental changes. (The word "consumers" instead of "fruit lovers" is used here very deliberately. Many of the changes have nothing to do with the joy of eating a great piece of fruit.) One thing consumers like is red — lots of red. Peaches and nectarines used to be prized for a golden skin tone; now people are buying red, equating it with ripeness. (In fact, the high-red blush on many new varieties of stone fruit actually makes it harder to tell when the fruit is ready. The red comes on early, obscuring the background color, which is what really predicts quality.)

But red has a hold that is almost subconscious. There's a story told by those in the stone fruit industry about a marketing experiment. A tasting panel was given two nectarines: one a fairly tasteless red variety, the other a great-tasting gold. Sitting around, tasting and talking about the fruit, the consumers unanimously agreed that the gold was a much better nectarine and that was the one they would buy. Then, on the way out the door, the panelists were offered boxes of nectarines as a thank you. One held the preferred golden fruit, the other the red. To a person, the consumers picked the red fruit to take home. Red sells.

That's not all. Although it seems impossible to most true peach and nectarine lovers, consumers in general show a preference for fruit that is firm and nearly crisp, as opposed to melting. Breeders are working on that, too.

Probably the biggest change in the stone fruit world in the past decade, though, has been the color of the flesh of the fruit. White-fleshed peaches and nectarines, practically nonexistent commercially ten or fifteen years ago, now make up a substantial part of the harvest. In one three-year period in the mid-1990s, plantings of white-fleshed varieties increased by more than 350 percent. This was particularly notable for nectarines — more than a third of all the trees that were planted during that period were white-fleshed. White-fleshed fruit went from being so little known that it was all but ignored in the official statistics to so popular that it required its own category. In 1992 white-fleshed fruit accounted for less than

2 percent of the total harvest. By 2002 it accounted for roughly 20 percent of all peaches and nectarines.

What's ironic is that until the 1950s, most of the nectarines that were grown in California were white-fleshed. Nectarines were a very minor crop at that point, and consumers were more willing to buy a fruit that had the familiar golden color of a peach. Perhaps more important, those older varieties were soft in texture and, because of their white flesh, showed bruises almost immediately. "You could ship them about as far as an ice-cream cone," one old fruit grower remembered.

To take advantage of these stone fruit trends, California growers have recently begun marketing fruit labeled "Summerwhite." The varieties included in this category combine the desirable characteristics of high-colored skin, white flesh and firm texture, and they make up more than 15 percent of the annual harvest.

Although white-fleshed fruit is popular domestically, the main demand for it comes from Asia. At one time as much as 80 percent of the white nectarines harvested in California went overseas, primarily to Taiwan. Furthermore, Asian customers love fruit that tastes extremely sweet. As a result, most of the white-fleshed peaches and nectarines that were planted at that time were not the old varieties, which had a nice tang to offset the sweetness, but rather what are called "sub-acid" varieties, which never develop much acidity at all. These work much like the so-called sweet onions, which actually don't contain any more sugar than other onions but taste sweeter because they lack the balancing acidity. To lovers of old-fashioned fruit, these sub-acid varieties tend to taste simple — sort of like sucking on a sugar cube. But there is significant money beckoning growers toward sub-acids, and it's hard to resist that incentive. From 1998 to 2002, 30 percent of all nectarines planted and 20 percent of all peaches planted were white-fleshed sub-acid varieties.

Up to now I've been lumping peaches and nectarines together as if they were the same fruit. What is the difference between a peach and a nectarine? Surprisingly little, botanists say; just one gene does it. In fact, so closely related are the two that sometimes peach seeds will sprout a sport nectarine tree and vice versa. Technically, peaches are "pubescent," which means that they have hair on the surface. How much hair differs from variety to variety. Some, such as the longtime favorite Elberta family, are quite hirsute, which gives them a slightly bitter, almost tannic finish that their fans find a quite appealing contrast to all that aromatic, juicy flesh. Nectarines have no fuzz — their skin is completely smooth, like that of an apple. Although there are a lot of differences among individual varieties of both fruits, as a general rule nectarines have a slightly more acidic character with an almost lemony top note, while peaches tend to be muskier and richer in flavor. Practically speaking, this flavor difference is of interest mainly to connoisseurs, and the two fruits usually can be used interchangeably.

The biggest challenge with both is finding fruit that is truly well matured. Peaches and nectarines are climacteric fruits, which means that they continue to ripen after being picked. You can buy hard peaches at the grocery store, leave them on the counter for a couple of days and wind up with some pretty killer fruit.

Maturity is another matter entirely. Although peaches and nectarines do soften and become juicier and more aromatic after harvest, they don't get any sweeter. That requires picking the fruit at the highest possible maturity. And though picking ripe fruit is no problem — you can smell it several feet away — choosing fruit that has been grown to maturity is tricky indeed. The best hint is the color of the fruit. This doesn't mean picking the peach that is the reddest — remember that blush is a genetic variation that has nothing to do with either ripeness or maturity. Instead, it means paying attention to the *quality* of the background color of the fruit. This is more difficult than simply deciding "yellow" or "green." Peaches and nectarines that have the most sugar and are the most mature have a background color (yellow) with a golden, almost orange cast. When you see a piece of fruit like this, pick it no matter what the variety is.

WHERE THEY'RE GROWN: Almost all of the nectarines grown in the United States come from California. Peaches are much more widely grown, with twenty-nine states harvesting significant amounts. But California still grows more than half — eight times as many as the next-closest states, Georgia and South Carolina.

HOW TO CHOOSE: Check the background color. Ripe fruit will be golden, not green. Mature fruit that hung on the tree long enough to develop the sugar will have a distinctive orange cast. Always with peaches and nectarines, trust your nose: fruit that is ripe and delicious will smell that way.

HOW TO STORE: If you buy fruit that is too firm, leave it at room temperature. Only when it begins to ripen should you move it to the refrigerator. In fact, chilling underripe fruit is about the worst thing you can do: it will turn the flesh mealy and dry.

HOW TO PREPARE: Nectarines don't need peeling. Peaches should be peeled before cooking; otherwise the skin will slip away into the dish on its own. To peel a peach, cut a shallow X in its blossom end, then blanch it quickly in boiling water. Rescue it to a bowl of ice water to stop the cooking, then peel away the skin with your fingers. The blanching time required will vary depending on the ripeness of the peach (in fact, you can peel very ripe peaches without blanching at all). To remove the pits from peaches and nectarines, cut the fruit in half lengthwise, following the cleft that runs down one side. Rotate the halves in opposite directions to free the pit. Peaches and nectarines are subject to enzymatic browning. If you are cutting them up in advance, sugar them to delay this reaction.

ONE SIMPLE DISH: Is there a better, simpler summer dessert than a perfectly ripe peach? Okay, serve it with shortbread. Or peel and slice peaches and marinate them briefly in a bowl of lightly sweetened red wine.

Peach Gelato

I learned this technique, which eliminates the need for an ice-cream machine, from the Sicilian chef Ciccio Sultano, who was singled out by the Italian magazine *Gambero Rosso* as one of Italy's great young chefs. It couldn't be easier, but the fresh peach flavor is astonishing. The texture should be somewhere between soft-serve ice cream and dense, chewy traditional gelato. Depending on the sweetness of your fruit, you may want to add more sugar.

8 SERVINGS

- 3 pounds peaches, peeled and pitted
- ¼ cup sugar, or more to taste
- ½ cup mascarpone, crème fraîche or yogurt

Cut the peaches into very small pieces. The smaller you cut them, the faster they will freeze and the finer the final texture will be. Arrange the peach pieces in a single layer on a baking sheet and freeze solid, about 2 hours.

Put the frozen peach pieces in a food processor with the sugar and grind briefly. Add the mascarpone and pulse until the mixture is smooth.

Empty the food processor into a small container and freeze again for 20 to 30 minutes before serving. If the gelato freezes solid, simply process it briefly again before serving.

Nectarine-Cardamom Ice Cream

Cardamom is used in all kinds of baked goods, particularly in Scandinavia. But it is rarely combined with peaches or nectarines. Still, its slightly acidic, flowery aroma lends a subtle note to this ice cream. Don't be tempted to add too much — it should be a suggestion rather than "Wow, that's cardamom!"

MAKES 1 QUART

- 2 **pounds nectarines**
- ¾ **cup sugar**
- 1 **tablespoon whole cardamom pods**
- ½ **cup whole milk**
- ½ **cup heavy cream**
- ½ **teaspoon salt**

Pit and chop the nectarines. You should have about 4 cups. Place them in a large bowl with the sugar. Stir to combine. Set aside for 30 minutes to macerate, stirring occasionally.

Lightly crush the cardamom pods with the base of a small saucepan, just enough to expose the little black seeds inside. Bring the cardamom, milk, cream and salt to a simmer in a small saucepan. As soon as bubbles appear around the rim and a skin forms on top, remove the pan from the heat and cover with a lid. Set aside for 30 minutes to steep.

Place the fruit in a blender and strain the cream mixture over the top, discarding the cardamom. Puree until smooth. The mixture will be slightly thinner than a milk shake. Pour into a bowl and refrigerate for 30 minutes.

Freeze in an ice-cream machine according to the manufacturer's instructions until the mixture has the texture of soft-serve ice cream. Spoon it into a container, cover and place in the freezer for 1 hour. Serve immediately. (If the ice cream is left in the freezer overnight, it may become icy. Allow it to warm slightly outside the freezer before serving.)

Nectarines and Blackberries in Rose Geranium Syrup

The subtle fragrance of the rose geranium leaves brings out a floral quality in the fruit. This is a great way to improve the flavor of ordinary supermarket nectarines.

6 SERVINGS

- 1 **cup water**
- ½ **cup sugar**
- 2 **tablespoons chopped fresh rose geranium leaves**
- 4 **nectarines, pitted and sliced into ¾-inch wedges (about 2½ cups)**
- 1 **cup blackberries**

In a small saucepan, whisk the water and sugar over high heat until the sugar is in suspension and no longer mounded on the bottom of the pan. Bring the mixture to a boil and cook until all the sugar is dissolved, about 5 minutes. If this happens before the water boils, bring to a boil anyway.

Remove the pan from the heat and add the rose geranium leaves. Let steep for at least 10 minutes.

Combine the nectarines and blackberries in a large bowl or divide evenly among six small ones. Ladle the warm syrup through a strainer over the fruit and serve. (The dish can be made up to 2 hours in advance and held at room temperature.)

FLAVORED SYRUPS

Nothing in the pastry chef's art can compare to a perfect peach — the melting texture, the heady perfume, the complex interplay of sweet and tart. But just how many times in a summer do you find perfection? Face it: most of the fruit we buy can use a little help. That's why desserts were invented in the first place. An assist can come in the form of something as tricky as puff pastry or as basic as a simple syrup.

Few things are easier to make than a simple syrup — just boil sugar and liquid until clear — and few allow as much room for experimentation. One of my favorite summer desserts uses a simple syrup scented with mint and lime zest as a sauce for thinly sliced melon — just five ingredients, but you would not believe the complexity of flavor.

Playing with an assortment of flavoring ingredients, I came up with some surprising syrups perfect for giving a discreet boost to summer fruit while you're waiting to find that perfect peach. Follow the directions in the recipe on page 216, substituting the following herbs, spices and fruits.

HERB/SPICE	FRUIT
Basil (2 tablespoons chopped fresh leaves)	Cherries, strawberries, figs
Black peppercorns (1 teaspoon crushed)	Cherries, figs, blackberries
Chamomile (½ teaspoon tea)	Cherries, blackberries
Jasmine (1 teaspoon tea)	Cherries, figs, blackberries
Lemon verbena (2 tablespoons chopped fresh leaves)	Figs
Rose geranium (2 tablespoons chopped fresh leaves)	Nectarines, peaches

Plums

· · · · · · ·

Whereas different varieties of peaches and nectarines can be so similar in appearance that even the farmer who raised them can't tell one type from another, there is no such difficulty with plums. Unlike their conformist stone fruit cousins, which come in standard uniforms of gold and red, plums stand out by flaunting their individuality like so many fashion models turned loose in a corporate headquarters. There are yellow plums, green plums, red plums, scarlet plums, purple plums and even plums that are almost black. This abundance has not escaped the notice of fruit marketers. Every summer grocery store produce managers across the country go a little crazy for a couple of weeks in what has come to be called in the trade "Plum-a-Rama" — piling up as many different colors of plums as they can find. And, in fact, it has been shown that offering a variety of colors results in far more sales than if the stores had just one.

Unfortunately, that variety is often scarcely more than skin-deep. Bite into most commercial plums today, and you'll find only slight variation on the theme of sweet and tart. The surprise that comes with the unexpected flavor notes found in great plum varieties — the wild herbaceousness of an Elephant Heart, the golden honeyed tang of a Wickson, the almost unbearable sweetness of a greengage — is increasingly elusive. For the most part, the market is dominated by large black plums that, in the carefully couched language of fruit catalogs, "can be good when fully ripe." That may be changing, though.

Plums are wildly promiscuous fruits, cross-pollinating and sporting with abandon. Plant breeders have taken advantage of this tendency by crossing varieties back and forth to try to develop improved varieties. The plum assortment in your neighborhood grocery store is likely to be a kind of living museum of the modern history of fruit breeding, with varieties developed by the great Luther Burbank at the turn of the twentieth century piled right next to varieties that were introduced only a couple years ago.

In his long career, Burbank developed more than eight hundred varieties of vegetables, fruits and flowers, ranging from the baking potato and the freestone peach to the Shasta daisy, but he was especially interested in plums, creating more than one hundred varieties. Probably his greatest was the Santa Rosa, named after the Sonoma County town in which he did most of his work.

The Santa Rosa is a dark red plum with amber flesh and a rich, winey flavor. Unfortunately, because it rarely grows to a very great size, it has fallen from favor. In the 1960s Santa Rosas dominated the plum harvest, accounting for more than a third of the total production. As recently as the mid-1990s, they were still as much as 10 percent of the harvest. Today they make up less than 1 percent. Still, it is a noisy minority — Santa Rosa is one of the few plum varieties sold by name in grocery stores.

Right next to Burbank's plums, you'll find cutting-edge crosses developed by Floyd Zaiger, Burbank's modern-day counterpart. Zaiger has revolutionized the plum business by crossing plums and apricots to come up with the Pluot (he owns the trademark on the name). Burbank had tried to cross a plum and an apricot, but his plumcot varieties were all either poor quality or light bearing. Plums account for the majority of the lineage in these crosses. Zaiger also has trademarked the Aprium, which refers to similar crosses where apricot characteristics dominate.

The first commercial Pluot was introduced in 1989, and those plum-apricot crosses now account for as much as a quarter of the plum harvest. The best varieties are Dapple Dandy (pale maroon skin and creamy red and white flesh, sometimes sold as Dinosaur Egg), Flavor King (reddish purple skin and red flesh) and Flavor Supreme (greenish maroon skin and red flesh).

WHERE THEY'RE GROWN: For all intents and purposes, California grows all of the country's plums, almost entirely in the same part of the Central Valley around Fresno where peaches and nectarines are grown.

HOW TO CHOOSE: Plums should be deeply colored, shiny and firm, but not hard (of course, neither should they be spongy). Don't worry if there is what looks like a white powder covering the surface; this is a natural "bloom."

HOW TO STORE: Like other stone fruits, unripe plums will improve if left at room temperature for a couple of days. And also like other stone fruits, if unripe plums are refrigerated, the quality will suffer. Once they're ripe, refrigerate them, tightly wrapped.

HOW TO PREPARE: Split a plum along the cleft that runs from stem to blossom end. The pit should pop right out.

ONE SIMPLE DISH: Simmer 1 cup red wine, ⅓ cup sugar and a sachet containing 4 whole cloves, 1 teaspoon black peppercorns and 1 cinnamon stick. When the mixture is clear and fragrant, add 1 pound pitted and quartered plums. Simmer until they soften a little, then refrigerate until chilled. Remove the sachet and serve over vanilla ice cream.

Spiced Plum Ice Cream

Plums are frequently described as "spicy," but until you taste this ice cream, with its subtle hints of pepper and allspice, you probably won't know just how appropriate that word is.

MAKES 1 QUART

- 2 pounds red-skinned plums or Pluots
- ¾ cup sugar
- ¾ teaspoon black peppercorns
- ½ teaspoon whole allspice
- ¾ cup heavy cream
- ¼ cup whole milk
- ½ teaspoon salt

Pit and chop the plums; you should have about 4 cups. Place them in a large bowl with the sugar. Stir to combine. Set aside for 30 minutes to macerate, stirring occasionally.

Crush the peppercorns and allspice with a heavy pan or with 1 or 2 pulses in a spice grinder. Leave in large pieces so they can be strained out later. Bring the spice mixture, cream, milk and salt to a simmer in a small saucepan. As soon as bubbles appear around the edge and a skin forms on top, remove the pan from the heat and cover. Set aside for 30 minutes to steep.

Place the fruit in a blender and strain the cream mixture over the top, discarding the spices. Puree until smooth. The mixture will be slightly thinner than a milk shake. Pour into a bowl and refrigerate for 30 minutes.

Freeze in an ice-cream machine according to the manufacturer's instructions until the mixture has the texture of soft-serve ice cream. Spoon it into a container, cover and place in the freezer for 1 hour. Serve immediately. (If the ice cream is left in the freezer overnight, it may become icy. Allow it to warm slightly outside the freezer before serving.)

Cornmeal Buckle with Plums

A buckle is an old-fashioned American dessert that is somewhere between a crisp and a cake.

8 SERVINGS

Topping
- ½ cup sugar
- 6 tablespoons all-purpose flour
- ¼ teaspoon salt
- 8 tablespoons (1 stick) unsalted butter, cut into 1-tablespoon pieces

Cornmeal Buckle
- 1½ cups all-purpose flour
- ¼ cup yellow cornmeal
- 2 teaspoons baking powder
- ½ teaspoon salt
- 8 tablespoons (1 stick) unsalted butter, softened
- 1 cup sugar
- 1 large egg
- ½ cup milk
- 1 pound plums, pitted and cut up

For the topping: In a food processor, pulse together the sugar, flour, salt and butter until the mixture resembles coarse crumbs.

For the buckle: Heat the oven to 350 degrees. Butter a 9-inch glass pie plate. In a large bowl, sift together the dry ingredients.

In another large bowl, with an electric mixer, beat the butter with the sugar and egg until the mixture is fluffy, 2 to 3 minutes. Add half the milk and beat until smooth. Gradually beat in the rest of the milk.

Stir the dry ingredients into the wet ingredients just until well moistened. The mixture will be the texture of cake batter. Fold in the plum pieces. Pour into the pie plate and spread evenly. Scatter the topping mixture evenly over the top. Bake until the top is golden brown and a toothpick inserted in the center comes out clean, about 45 minutes. Serve warm.

Growers and Global Competition

REINVENTING THE TOMATO

For what seemed like forever, the tomato was the thing most people pointed to when they wanted an example of all that had gone wrong with modern agriculture and the supermarket produce section. It was cottony and pink and had no flavor. Have you noticed that hardly anyone is saying that anymore?

This is not to suggest that the tomatoes at your neighborhood supermarket have suddenly gotten as good as the ones in your backyard (or at your local farmers' market, or even at a high-end grocery), but remarkable progress has been made — at least in variety if not always in quality. Walk into your produce department today, particularly during the prime summer growing season, and you might find dozens of different kinds, some of them varieties that were available only to die-hard collectors just a few years ago.

What has happened is nothing more mysterious than the "invisible hand" of capitalism invoked by eighteenth-century economist Adam Smith. But ironically, whereas Smith was talking about how the preference for domestic goods indirectly benefited the consumer by strengthening local producers, in the case of tomatoes, it was a lust for imports that turned American farmers around.

Let's back up a minute to those bad old days in the early 1990s. Back then, tomatoes basically came in two types: "mature-green" and "vine-ripe." Those two classes made up more than 90 percent of the U.S. tomato harvest, although there were small amounts of other types as well: cherry tomatoes usually, and if you lived near an ethnic or upscale market, you could find Romas, or plums, too. In the summer, stores might promote something they called a beef-steak. Nobody liked any of them very much, but almost everyone accepted the situation as the way things were. If you wanted a good tomato, you had to grow your own.

One group that wasn't complaining was the tomato farmers, who were doing a very good business. In 1994 fresh tomatoes were almost a $1 billion industry, and the farmers believed they had it all figured out. They had a good run of plants that resisted most common pests. They picked their fruit rock hard so it was practically indestructible. (At the time, a news photo of an overturned tomato truck showed a lot of twisted metal but most of the fruit intact.) When they needed to sell some tomatoes, they would gas them to some semblance of red. And best of all, they had a guaranteed market: it is almost unthinkable that a fast-food hamburger or taco would not have some tomato on it. In fact, the food service industry uses roughly half of all the fresh tomatoes grown in the United States.

To get a full understanding of the situation as it then stood it is worthwhile to linger for a moment on the subject of "mature-green" and "vine-ripe." Although it might not be apparent to the consumer (since neither type tastes much like the tomato we remember), there's more to the difference between them than simply the color at which they are picked. They actually come from different types of plants, are grown in different ways and are sometimes sold to different people.

Even today mature-green tomatoes make up the bulk of the fresh tomato harvest — somewhere around 75 percent. They are picked from plants that are called "determinate," which means the vines grow only to a certain point and then stop. Determinate tomatoes grow on bushy plants that sprawl whichever way they will. Farmers don't need to worry about training them because they'll only get so big. Because of the way the plants are structured, harvesting them is a bit of a chore: the pickers have to sort through the foliage to find the fruit. This means that the fields are harvested in only one or two sweeps. Whenever a certain percentage of the tomatoes starts to show some color, almost everything gets picked. And when growers refer to "showing color," they're not seeing red. The first pale blush of green to cream is enough. As a result of this mass approach to harvesting, a good many of the tomatoes are significantly underripe when they are picked. This is not much of a problem for fast-food places, where tomatoes are used mainly as a source of symbolic color. And it even works for consumers in the winter, when most of us are so desperate for tomatoes that we'll take almost anything we can get. The majority of these tomatoes are grown in Florida, which is about the only place that can grow tomatoes outdoors at that time of year. During the summer, when it's too hot to grow in Florida, mature-green tomatoes come from the Central Valley of California.

Vine-ripe tomatoes come from plants that are called "indeterminate," which means they will keep growing and producing tomatoes as long as there is sunlight and warmth. As a result, farmers need to stake them in place and keep them trained. The extra labor means that the tomatoes must be sold for a higher price, so they are picked a few days later than they would be if they were mature-green.

This improvement isn't as dramatic as the name "vine-ripe" might imply. Usually they are harvested at what is called the "breaker" stage, when the fruit just begins to show some red (technically, not more than 10 percent of its surface). Still, a tomato matures quickly at this point in its life, and there can be a discernible difference in flavor. Also, because the plants are staked and trained, the fruit is easier to pick, so the harvest proceeds gradually rather than in one giant sweep. It's not unusual for a vine-ripe field to be picked every other day; therefore, a higher percentage of the harvest will be ripe than with determinate tomatoes.

Because the plants are producing for such a long period of time, they need a fairly mild temperature. Most of the vine-ripe tomatoes grown in the United States come from the coastal regions of California, although vine-ripe tomatoes are among the most geographically dispersed crops, grown in thirty-five states.

As sweet as tomato growers might have thought they had it,

nothing lasts forever. In the mid-1990s, the tomato world was turned upside down when imported fruit hit the American market with a bang. The shocking thing was where these revolutionary new imports came from. Imported tomatoes had long been a significant part of the American scene, particularly in the winter, but historically most imports came from Mexico, where a thriving tomato industry is located in the state of Sinaloa. (Imports from Mexico usually come close to equaling the individual harvest of either Florida or California.) The Mexicans had particularly good luck in the early 1990s with special breeds of tomatoes that could be picked a little later than mature-greens, so they had somewhat more flavor but still a long shelf life.

What took American farmers by surprise were the tomatoes coming from a most unexpected place: Holland. In the mid-1990s Dutch tomato exports to the United States skyrocketed, increasing more than 800 percent in only a couple of years. What was truly revolutionary was that these tomatoes didn't fit the old mold. Rather than sending over another variation on the old mature-green/vine-ripe model as the Mexicans had done, Dutch farmers practically reinvented the tomato. As a result, Holland went from nowheresville in the tomato world to becoming the second-largest exporter to the United States.

Customers not only accepted this Dutch fruit; they actively sought it out. While American fresh tomato prices at wholesale hovered around 25 cents a pound, Dutch tomatoes averaged 80 cents. The effect on American growers was immediate and drastic. From 1992 to 1995 the wholesale price of domestic tomatoes dropped for three straight years, falling a total of almost one third. Within five years tomato imports tripled and domestic production declined by 20 percent. Florida, which had concentrated on winter-grown mature-greens, had to cut back its harvest by 40 percent.

Grown in hothouses, these new tomatoes could be lavished with the kind of care that resulted in fruit with something approaching real flavor. And they came in more colors and shapes than the typical red and round. Using tomato varieties developed primarily by Israeli breeders, the Dutch supplied American shoppers with squat tomatoes, yellow tomatoes and pear-shaped tomatoes. Most important (commercially anyway), some of these tomatoes were sold still clustered on the vine. The tomatoes themselves might not actually have had more flavor than other tomatoes, but they sure smelled as if they did to shoppers. Tomato greens are extremely aromatic — maybe even more so than the fruit — and much of what we remember as a fresh tomato's perfume is actually the smell of the vines and leaves.

Developed in Italy and enthusiastically embraced by Dutch greenhouse growers, tomatoes on the vine (known in the industry as TOV) became an important category within only a few years. By 1999 tomatoes on the vine accounted for 13 percent of all greenhouse tomatoes sold in the United States, and by 2003 they represented almost a quarter.

Although this slight increase in variety hardly seems revolutionary today, it was a grand start considering the time and place. Produce managers quickly found that offering an assortment of tomatoes increased the sales not just of the new types but of the old ones as well. Consumers, it seems, like to have options — even if they often end up buying the same old thing. The produce section with the greatest assortment of products is the one that is judged the best. This started a veritable tomato arms race among high-end supermarkets. Grocery stores that not long before had carried three or four types of tomatoes suddenly were carrying eight or nine. And some ultra-ambitious retailers were advertising as many as twenty-two different types of tomatoes.

As radical as this seemed, it was really only the first step. In 1997 hothouse tomatoes — buzz-worthy though they were — generated only about 7 percent of retail tomato sales in the United States. And considering the cost of transporting fresh produce from Holland, it was unlikely that Dutch tomatoes would ever be more than an attractive niche item.

But then came another shock: more imports from another totally unexpected quarter. Canadian growers started adapting the Dutch tricks, and from a much more manageable distance. The idea that a country widely regarded by Americans as the frozen north could excel at growing tomatoes seemed even more far-fetched than that the Dutch could. But the Canadian greenhouse tomato industry proved to be an even bigger threat. From almost nothing in the early 1990s, Canadian tomato exports to the United States increased 600 percent by 2003, eventually accounting for 17 percent of all American tomato sales and a whopping 37 percent of all American tomato sales at retail. (Those long-lasting, easy-slicing mature-greens continue to dominate food service, which still accounts for half of all tomatoes sold.) Wholesale prices for domestic tomatoes, which had started to improve (or at least stabilize) after the Dutch invasion, tumbled again, falling two out of three years from 1999 to 2001 and losing a net 10 percent.

These two invasions threw American farmers into a tizzy. Fresh tomato production overall has grown only 8 percent since 1990, and in Florida it has actually declined since the introduction of hothouse tomatoes. Even in California, where high-quality vine-ripes resisted competition somewhat better, growers were forced to rethink how they did business. Because of the extremely high costs of starting up greenhouses in the United States (the price of establishing a greenhouse full of tomatoes runs to more than $1 million per acre, as opposed to around $3,000 for field tomatoes), the hothouse option has been pretty much off the table. There are only four large growers of hothouse tomatoes in the United States, and the industry as a whole has been racked by financial uncertainty.

In Florida, some growers shifted to other varieties. Mature-greens dropped from more than 85 percent of the total harvest in 1997 to less than 75 percent in 2003. Tomatoes other than mature-greens and vine-ripes went from almost nothing in 1997 to more than 15 percent of the total harvest in 2003. Other growers adapted by simply picking their mature-greens later. More than 10 percent of Florida's mature-greens were picked at the vine-ripe stage in 2003. In California, the share of mature-greens declined from more than three quarters to just over two thirds.

Some adventurous growers began exploring other tomato options, including heirloom varieties that not long ago were found only in the gardens of passionate collectors who saved seeds at the end of each season to share among themselves.

Heirloom tomatoes such as Brandywine, Cherokee Purple and Radiator Charlie's Mortgage Lifter were the results of hundreds of chance mutations and eons of human migrations. They came from Italy and France, Spain and Poland, even Russia. They were everything mainstream tomatoes were not — the kind of fruit that made their fans think smugly, *You won't find* that *in the supermarket.* By definition, heirloom tomatoes are open-pollinated, as opposed to hybrid, so the seeds grow true to the parents. (Although there is no formal age limit, it is generally agreed that an heirloom variety must have been around for at least three generations to qualify.)

You can usually recognize heirloom tomatoes first by their imperfections. They are plainly and outspokenly old-fashioned. They tend to have unusual shapes and frequently odd colors. They wear their wrinkles and blemishes as signs of character. Amid the perfect uniformity of the modern produce section, they stick out like the Queen Mum at a fashion shoot.

But their appeal is undeniable — and profitable. When most tomatoes sold for less than $2 a pound, heirlooms went for as much as $6. And even at those prices, there was no shortage of buyers. At some high-end groceries, heirloom tomatoes became the single best-selling summer produce item in terms of dollar volume. Ironically, most of these heirlooms had at one time or another been discarded by commercial growers because of their cosmetic blemishes, thought to discourage shoppers, and because they have much thinner skins that puncture easily, leading to rapid spoilage.

First the heirlooms caught on at farmers' markets. When chefs snapped them up and bought as many as they could, the cult of the heirloom tomato was born. Suddenly, every tomato in every fancy restaurant had a provenance that was spelled out in excruciating detail on the menu. High-end produce managers weren't sleeping. They began to seek them out, too. To be sure, these tomatoes rated hardly a blip on the radar screen of the commercial tomato world.

No one keeps statistics on them, but today there are almost certainly fewer than 300 acres of heirloom tomatoes grown in all of California (as opposed to more than 30,000 acres of fresh tomatoes total). Still, they have had an influence that far outweighed their actual dollar worth. Not only did they reinforce the message to produce marketers that the old variety paradigms weren't working, but they also prompted some supermarket chains to change their distribution patterns.

The industry standard is to have all fresh tomatoes delivered to a central warehouse, where they are sorted and from which they are shipped to individual stores, a process that can take two or three days. (This is for upscale groceries. Most tomatoes go through even more hands between farm and market.) Because heirloom tomatoes were such a premium product, some markets began allowing individual farmers to deliver their tomatoes directly to the stores. This allows an extra two or three days of ripening before harvest, which can result in an immense improvement in flavor. The heirloom boom also further blurred the line between farmers' markets and supermarkets, as many growers supplied both. Today, during the peak summer months, it's not uncommon to find heirloom tomatoes with the name of your favorite farmers' market grower both in specialty markets and at supermarkets clear across the country.

Although it is good to see these old varieties finding new popularity, many hard-core tomato fans question whether something vital is being lost in the translation. Is an old tomato variety grown, picked and packed by modern commercial standards really worth celebrating? Frequently, the answer is no. Tomatoes are grown, not manufactured. They are not Fords (or even Cadillacs), and there is more to good quality than a brand name, no matter if it is Brandywine or Radiator Charlie's Mortgage Lifter. To make the long journey to market, in many cases these heirlooms were being picked at the vine-ripe stage or even earlier. Too often these were heirloom tomatoes in name only.

The next great stage in tomato evolution delivered much better flavor on a consistent basis. Called grapes, or miniatures, these tiny tomatoes, about the size of the tip of your little finger, are amazingly sweet. Best of all for the growers, they have a thick skin that not only pops when you bite into them but also protects them well enough that they can be harvested at nearly full ripeness. Although they resemble the little cherry tomatoes that have been with us forever, these new tomatoes are something different.

The first grape tomatoes — which are a little smaller than cherry tomatoes and more oblong than round — began hitting store shelves on the East Coast in 1997. One of the truly radical things about these miniature tomatoes was that, compared to heirlooms, they followed the reverse path to popularity, starting out as a commercial product and then filtering down to small farmers and gardeners. Miniatures were introduced by a commercial grower in Florida named Andrew Chu, who had heard about them from a friend in Taiwan. He ordered the seeds from the Known-You Seed Company there and first planted them in 1996. In 1997 he began packaging the tomatoes in easily recognizable, clear plastic clamshell boxes and distributing them to supermarkets on the East Coast. When other growers tasted these tomatoes, they jumped on the bandwagon. When Chu tried to trademark the name "grape tomato" in 1998 to protect his market share, a series of legal battles ensued, which were eventually settled out of court.

All the while, the tomatoes were gaining in popularity. Between 1999 and 2003 the volume of grape and cherry tomatoes sold increased more than 300 percent, and their price — already higher than those of standard tomatoes — increased more than 400 percent over the same period. People weren't buying the little gems just because they were cute; they were buying them because they tasted better.

Most miniature tomatoes are naturally sweeter than their bigger brothers. Whereas regular tomatoes have a sugar content of 4 or 5 percent, cherry and grape tomatoes reliably sweeten into the 8 to 9 percent range, and sometimes even higher. Because they're so small, they ripen much more quickly than regular tomatoes — a real boon in cool, cloudy climates — and they last longer after picking than do other tomatoes, so they can be picked nearly dead ripe and still be delicious a week later when you get them home from the supermarket. In fact, one of the biggest hurdles tomato farmers faced was teaching crews how to pick them. Rather than harvesting tomatoes at the slightest sign of a blush, pickers needed to be taught to wait until the colors develop fully.

The original grape tomatoes belonged to a single red variety called Santa, but they now come in a dazzling assortment of colors and shapes. Some are round, some are grape-shaped, and some look like miniature pears. They are every color in the tomato rainbow: red, green, yellow, white, even purple-black. Sungold, a yellow, is well on its way to being an established favorite, as is Juliet, a red oval. Candy looks like a miniature German Pineapple tomato, and Tigerella is striped red and yellow. Taste an assortment of them, and you'll realize there is no such thing as one single "tomato flavor." Rather there is a spectrum, running from almost lemony to nearly beefy.

Fresh tomato consumption, which was less than 12½ pounds per person through the 1970s and early 1980s, now stands at more than 19 pounds. Total sales have increased to more than $1.3 billion — behind only lettuces in the commercial vegetable hierarchy. Even Florida, so battered by the initial flood of imports, has rebounded. And of the 1.5 billion pounds of tomatoes it grows every year, fully 7 percent are the high-flavor, high-profit grapes. Much more important for consumers, what once stood as a symbol of everything that had gone wrong with commercial agriculture now represents a promise of what can be accomplished with a return to the simple values of flavor and variety. Although Adam Smith would no doubt be dumbfounded at a world in which so much attention is paid to a single fruit (and so much money can be made from it), it's hard to imagine that he would not be thrilled by the result.

Fall

Broccoli and Cauliflower

. . . .

Anyone can tell the difference between broccoli and cauliflower, right? Broccoli is green; cauliflower is white. Broccoli comes on long branching stems; cauliflower has a stalk so short you can barely see it. Well, yes and no. The family tree is as tangled and confusing as European royalty, and there's no *Debrett's Peerage* to help you sort things out. In fact, so tangled is the lineage that scientists are still scrambling to determine exactly who is a broccoli and who is a cauliflower.

Both broccoli and cauliflower are domesticated versions of a wild cabbage that grows in coastal regions of Europe. As such, they share a common genus and species: *Brassica oleracea*. Generally, broccoli (*B. oleracea italica*) is typified by its massive collections of immature flower buds that appear atop one central stalk as well as several side shoots. Cauliflower (*B. oleracea botrytis*) is recognized by its short central stem and even bigger collection of what botanists describe as "aborted" flower buds — "aborted" because fewer than 10 percent of them will ever flower.

So far so good — until you begin to explore outside the United States or become curious about any of the many plants that fall somewhere in between those two extremes. In Italy, for example, the plant they call broccoli is much closer in appearance to what we call cauliflower. Certainly, it is green, but a much paler shade (there is even a white variant called *broccoli bianco*), and its branch structure is short and squat. Indeed, the branching vegetable we call broccoli is much closer to what the Italians call *broccoletti*, which is not to be confused with the new American variety we call Broccolini.

And then there are all of the broccoli-cauliflower intermediary cousins. You may have seen one called broccoflower. It has been tried repeatedly but never seems to catch on. This is actually much closer in appearance to what the Italians call *broccoli Calabrese*, with a pale green head made of tightly bunched buds. There is also the oddly shaped Romanesco, with its buds bunched in tight cones, looking like a vegetable designed by Gaudí. Cauliflower has its share of oddities as well, including one with a purple head. So confusing is it all that several years ago a team of British researchers who tried to sort it out determined that what is usually called purple cauliflower is actually a broccoli and that Romanesco is actually a cauliflower. Current work is focused on examining the plants' DNA.

Cooks don't need DNA to tell the difference; we can just take a taste. Broccoli is green, wild and assertive. Cauliflower is nutty, subtle and quiet. The taste of broccoli will come through no matter what other ingredients you throw at it, but it will shrink from extended cooking. Cauliflower's flavor is more malleable and seems to deepen and become richer the longer it cooks. In part this difference in character is directly traceable to their respective amounts of certain chemicals that, when the vegetables are cooked for very long, form sulfur compounds. Broccoli is up at the top of the sulfur ranking — right behind cabbage — but cauliflower is way down at the bottom. Combine that with the way chlorophyll-rich broccoli changes from green to olive drab with overcooking, and you have two vegetables that, though very closely related genetically, need to be treated very differently.

Broccoli should be cooked quickly and used in ways that show off its bright color, verdant flavor and crisp texture. Use broccoli in salads (steamed or blanched first, please; you're not running a steak house salad bar), as a side dish or in pasta toppings. Broccoli is one of the great flavor matches for olive oil and lemon, and, truth be told, that simple treatment is difficult to beat. You can make it a bit more complicated if you like: sauté minced garlic, red pepper flakes, capers or salted anchovies (or all of the above) in the oil first. Add toasted pine nuts or slivered olives, or the very Sicilian toasted bread crumbs, at the end. A salty sheep's or goat's milk cheese works well, too, such as pecorino Romano or feta.

Whereas broccoli dishes should be vibrant and immediate, cauliflower is a vegetable that repays more careful cooking. There is a world of flavor in cauliflower, if only you have the patience to discover it. You can cook it quickly and use it in almost any wild way you'd use broccoli. Or you can cook it slowly until it is nearly melting in texture and transform it into one of the most elegant vegetables on the planet. Prepared this way, cauliflower is wonderful combined with butter and cream. Bake it in a custard, and you get a dish of almost shivering delicacy, with a deep, profoundly earthy flavor. This is one of fall's most regal vegetable creations, and it's an amazing base for pairing with some flavors that might surprise you. Despite its down-market image, cauliflower is amazingly good with caviar and may be even better with white truffles. (The perfume of white truffles is formed by chemical compounds related to some found in cauliflower.)

To get that really rich, deep cauliflower flavor for a custard, you need to cook it thoroughly. The timing is important: the longer you cook it, the more powerful the mustardy flavors will be. You want them to be pronounced enough that they will cut through the cream, cheese and eggs, but not so strong that they become obnoxious. When the cauliflower is done, it should be just soft enough that you can crush a floret between your fingers. You can boil cauliflower, but steaming is better. With boiling, the cauliflower taste is muted. The vegetable exchanges liquid with the cooking water, losing flavor and picking up only a little saltiness in exchange. (You do salt your cooking water, don't you?) Steaming keeps more of the taste intact because the vegetable never touches the liquid.

What about all of those confusing "intermediate" brassicas, the ones between broccoli and cauliflower? How do you prepare them? The best thing to do is assign them to one end of the spectrum or the other. Forget botany and use two simple criteria to do this: curd and color. Curd refers to the texture of the unopened flower buds in the crown. If it is truly "curdy" — tightly grained and almost sandy in texture — cook it like cauliflower. If the buds are distinct and separate, cook it like broccoli. Similarly, if the heads are dark green, they must be cooked like broccoli to avoid the "olive drab" syndrome. If they are lighter in color — white or pale yellow — you can safely cook them for an extended period. The various bunching broccoli types — broccoli rabe (also called raab, rape and rapini, depending on who is selling it) and the new one alternately called Broccolini and Asparation — definitely take brief cooking (and, because of their bitterness, are brilliant with sausages). The curious-looking Romanesco can take longer cooking if it is not a too-dark green (some varieties are).

Once scorned by the general public, broccoli and cauliflower have been enjoying dramatically increasing popularity (and prices) over the past several years. The value of the cauliflower crop in the United States increased almost 40 percent between 2001 and 2003, and that of broccoli increased almost 30 percent. Even so, the United States still grows more than twice as much broccoli as cauliflower. Although the acreage planted has remained fairly constant, there has been a boom in the sales of precut cauliflower and particularly of precut broccoli.

Broccoli

· · · · · · ·

WHERE IT'S GROWN: Almost all of the broccoli grown in the United States comes from California. A cool-weather crop, it is harvested in the southern Imperial Valley during the fall and winter and then moves progressively north to Salinas in the summer. A small amount is imported, mostly from Mexico. Broccoli is in the market year-round, but its prime season is from late fall to early winter.

HOW TO CHOOSE: There are two keys to picking broccoli. Remember that the flower buds are supposed to be immature.

Reject any broccoli that shows any signs of little yellow flowers. (The flower buds are also where decay will begin. Look carefully for any soft or slimy spots.) Also check the stalk: try to pierce it with your thumbnail. As broccoli matures, some of its sugars are converted to lignin, a woody material. Broccoli that was picked too late will be tough and lack sweetness.

HOW TO STORE: Broccoli is one of the most sensitive vegetables you can buy; it spoils very quickly. Try to use it the same day you buy it. Failing that, treat it like lettuce — keep it tightly wrapped in the crisper drawer.

HOW TO PREPARE: Separate the heads from the stalk because they cook at very different rates. But don't discard the stalk, even if it's a little woody. Peel it down to the pale core, and it's delicious. Giving broccoli that's a little tired a 30-minute soak in ice water will help crisp it.

ONE SIMPLE DISH: Blanch broccoli heads until they are just tender, then chill them. Just before serving, make a vinaigrette with olive oil, fresh lemon juice and a little minced garlic. Toss the broccoli heads in the vinaigrette and season with salt to taste. If you want a little fancier presentation, you can arrange the heads flower side down in a bowl and then invert it onto a serving plate.

Cauliflower

. . . .

WHERE IT'S GROWN: Almost all of the cauliflower grown in the United States comes from California, at the same times and from the same places as broccoli.

HOW TO CHOOSE: Cauliflower heads should be a pale creamy color. This is achieved by wrapping the heads with the leaves. Some varieties do this naturally; in others the leaves have to be tied in place by growers. Cauliflower heads that have not been carefully tended will show "sunburned" dark spots. If there are only a few, this is cosmetic. But if they cover a larger area, it can be an opening for decay. Reject any heads that show signs of physical damage, as cauliflower spoils quickly.

HOW TO STORE: Like broccoli, cauliflower is extremely perishable. Keep it tightly wrapped in the crisper drawer of the refrigerator. Try to use it the same day you buy it.

HOW TO PREPARE: Separate the heads from the stalk and break the heads into bite-size florets. The stalk can be chopped and cooked, but because it is usually so small, this isn't as important as with broccoli.

ONE SIMPLE DISH: Cauliflower is amazing roasted. Break it into florets and toss them with olive oil and salt and pepper to taste. Dump them into a baking pan and roast at 400 degrees until they're lightly browned and tender.

Broccoli Chopped Salad

This cold-weather salad has it all: crunchy broccoli; salty, peppery crisped pancetta; sharp green onions; toasted walnuts, and creamy, pungent blue cheese. The form and the flavors are familiar, but the combination is sophisticated.

6 SERVINGS

1¾ **pounds broccoli**
4 **slices pancetta (about 3 ounces)**
½ **pound white mushrooms**
⅓ **cup chopped green onions (about 3)**
¼ **pound Gorgonzola, crumbled**
¾ **cup chopped walnuts, toasted**
Olive oil
2 **tablespoons tarragon vinegar, plus more to taste**
Kosher salt
Freshly ground pepper

Remove the florets from the broccoli stalk and break any large ones into bite-size pieces. With a sharp paring knife, cut off the dried-out base of the stalk, then peel the tough skin, leaving it as a rough, squared-off log. Cut the stalk into ¼-inch dice.

Add water to the bottom of a steamer and bring to a boil. Place the diced stalk in the steamer basket and cook, covered, for about 1 minute. Add the florets, cover again and cook until the broccoli is bright green but still crunchy, another minute or two. Set aside.

Place the pancetta slices in a cold skillet and place over medium-low heat. Cook until well browned on one side, about 10 minutes, then turn and cook until the other side is browned, another 3 to 4 minutes. Do not rush the cooking, or the pancetta will not be crisp. When the pancetta is crisp and well browned, transfer to a paper towel to drain. Set the drippings aside to cool.

Remove the bases of the mushrooms and cut the mushrooms into quarters.

Place the broccoli in a large bowl and add the mushrooms, green onions, Gorgonzola and walnuts. Crumble the pancetta over the top.

Pour the pancetta drippings into a measuring cup. Add enough oil to make ½ cup, then pour the mixture into a small bowl. Add the vinegar and 1 teaspoon salt and whisk to emulsify.

Pour the dressing over the salad and stir to combine well. Taste and correct for seasoning with salt and vinegar. Finish with a good grinding of pepper. Refrigerate until ready to serve.

Pasta with Broccoli and Sausage

The trick with this recipe — if you can call it a trick — is getting the timing just right so that the pasta, broccoli stems and broccoli florets are perfectly cooked at exactly the right time. Don't worry: even if you miss by a bit, this dish will still be delicious.

4 SERVINGS

- 2 tablespoons olive oil
- 3 hot Italian sausages
- ½ pound broccoli
- 1 pound short pasta, such as gnocchi or penne
- 3 garlic cloves, thinly sliced
- Juice of ½ lemon
- 2 tablespoons minced fresh parsley
- 1 ounce Parmigiano-Reggiano, grated (about ¼ cup)
- Salt

Bring a large pot of generously salted water to a boil. While waiting for the water to boil, heat the olive oil in a large skillet over medium heat. Slit the sausages and squeeze the meat into the skillet; discard the skins. Break up the sausage with a wooden spoon and cook until the meat is in small pieces and deeply browned.

While the sausage is cooking, prepare the broccoli. Trim the florets from the stalk. Using a thin, sharp knife, peel the stalk and cut off the dried-out base. Dice the stalk into small pieces.

When the water comes to a boil, add the pasta. Keep track of the time. Add the sliced garlic to the sausage and keep warm over low heat. After the pasta has cooked for about 3 minutes, add the diced broccoli and stir to make sure the pasta isn't sticking. After 4 to 5 minutes, add the broccoli florets and cook just long enough to soften them slightly and brighten the color, about 2 minutes. At this point, the pasta should be done.

Add the lemon juice and minced parsley to the sausage, then drain the pasta and broccoli, reserving ¾ cup of the cooking water. Add the pasta, broccoli and reserved cooking water to the sausage and cook over high heat, tossing and stirring constantly to lightly coat the pasta with the juices as they reduce.

Add about half of the grated cheese and stir to combine well. Season with salt to taste. Divide among four heated pasta bowls and serve immediately, passing the remaining cheese at the table.

Cauliflower Custard

Mention cauliflower, and "silky" and "elegant" are probably not the first two adjectives that come to mind. But this recipe, which is both, demonstrates that there are no ingredients so ordinary that they can't be transformed into something special. It makes a superb first course for a special dinner party.

6 SERVINGS

14 ounces (about 4 cups) cauliflower florets
1¼ cups heavy cream
4 large eggs
1 teaspoon salt
   Freshly grated nutmeg
½ garlic clove
2 tablespoons butter
½ cup fresh bread crumbs

Heat the oven to 300 degrees. Steam the cauliflower until quite tender, 7 to 10 minutes. When it's done, you should be able to smash a floret between your fingers. Undercooking the cauliflower will make the custard grainy.

Reserve 4 florets for garnish. Transfer the remaining cauliflower to a blender and pulse several times. Add the heavy cream and puree until smooth. Pulse in the eggs, salt and nutmeg to taste.

Strain the puree into a large measuring cup. Stir the puree with a rubber spatula to help it flow through the strainer, but do not press it; pressing will also make the custard grainy. Divide the strained puree evenly among six ½-cup ovenproof ramekins.

Place the ramekins in a large roasting pan and place the roasting pan in the oven. Pour boiling water into the roasting pan to come halfway up the sides of the ramekins. Bake the custards until the center of each just barely jiggles when you shake the pan, 35 to 40 minutes.

While the custards are baking, prepare the topping. Mince the garlic. Heat a small skillet and melt 1 tablespoon of the butter over low heat.

Add the garlic and bread crumbs and cook until the crumbs turn golden brown, about 5 minutes. Remove from the heat and set aside. Slice each reserved cauliflower floret from top to bottom into 3 pieces. In a separate pan, melt the remaining 1 tablespoon butter. Cook the florets until golden brown, about 10 minutes. Remove and drain on paper towels. Set aside.

When the custards are cooked, remove from the water bath and let stand at room temperature for 10 minutes to set. Sprinkle the bread crumb mixture evenly over the top and garnish each custard with 2 cauliflower slices. Serve immediately.

Garlicky Braised Cauliflower with Capers

I can't even begin to count the number of times I serve this dish in the fall and winter. It is equally good with cauliflower or broccoli. Serve it as a side dish to complement a roast, or as a light main course, maybe with some ricotta salata shaved over the top. It even works well as a pasta sauce.

6 SERVINGS

- 3 tablespoons olive oil
- 4 salted anchovy fillets, rinsed and minced
- 3 garlic cloves, minced
- ¼ teaspoon red pepper flakes
- 1¼ pounds cauliflower florets
- ½ cup water
- ¼ cup chopped fresh parsley
- 3 tablespoons capers, with their liquid
- Salt

Combine the olive oil and anchovies in a skillet over medium heat. Cook, stirring, until the anchovies begin to melt into the oil, about 3 minutes. Add the garlic and red pepper flakes and keep cooking until the garlic softens, about another 3 minutes.

Add the cauliflower florets and water. Cover tightly and cook until the cauliflower becomes slightly tender — soft enough to be pierced with a knife, but not so soft that it can be crushed, about 7 minutes.

Remove the lid and raise the heat to high. Cook, stirring, until the water evaporates, leaving behind a thin layer of syrup in the bottom of the pan, about 5 minutes. Add the parsley and capers and cook briefly. Season with salt to taste and serve warm.

Mushrooms

. . . .

The little town of Kennett Square, Pennsylvania, hardly seems like an agricultural hotspot. In fact, the rolling hills around Kennett Square are not covered by farm fields at all. Tucked away in the southeastern corner of the state, near Maryland and Delaware, it is primarily known for its quaint architecture and as the hometown of Herb Pennock, an early baseball star who was inducted into the National Baseball Hall of Fame in 1948. Even so, farming is an important part of the local economy, but here in Kennett Square, all of the farming is done indoors. This town of 5,300 calls itself the Mushroom Capital of the World, producing more than half of all the domesticated fungi grown in this country.

The rise of Kennett Square as the mushroom capital was hardly inevitable. Cultivated mushrooms, after all, require no special soil or climate that would dictate one growing area over another. Rather, the town's ascension was due to a timely combination of happenstance and economic logic. The mushroom industry here dates to the turn of the twentieth century, when Kennett Square was already well established as a center for growing hothouse flowers for the floral trade (particularly carnations).

Kennett Square's mushroom heritage got its start when a Quaker farmer named William Swayne figured he could stack a second crop in his greenhouse and grow mushrooms in the damp, dark area underneath his flower benches. This would have been nothing more than a happy bit of trivia if it hadn't been for the town's central location, convenient to the major markets of Philadelphia, Pittsburgh, Buffalo, Baltimore, Washington, D.C., and New York. These cities provided the kind of educated, ethnically diverse consumers who would eat mushrooms, which at that time were not a popular item. Just as important, these cities were rich in the stuff that was necessary for mushrooms to grow — mainly horse manure, hay and straw. Trains bound for the big cities would roll out of Kennett Square loaded with mushrooms and return loaded with the collected detritus of those urban areas. By 1924, 85 percent of the mushrooms in the United States were grown in Pennsylvania, and by far most of them were grown in the Kennett Square area. Much has changed in the mushroom industry since then — notably, it is much less dependent on horses — but Kennett Square is still the center of mushroom propagation.

Dozens of varieties of mushrooms are commercially cultivated around the world, but more than 90 percent of those grown in the United States belong to one species — *Agaricus bisporus*. This familiar white mushroom was domesticated in Paris during the 1800s (in old texts you may find it referred to as *champignon de Paris*).

A brown variant of the same mushroom is sold as the crimini, and overgrown crimini are sold as portabellos. Fresh shiitakes (*Lentinus edodes*) are probably the next most common domesticated mushrooms, but there are also plenty of oyster mushrooms (*Pleurotus*) species around, as well as enoki (*Flammulina velutipes*). If you shop at Asian markets, you'll find an overgrown oyster called the king eryngii (*Pleurotus eryngii*); the maitake, or hen-of-the-woods (*Grifola frondosa*); and the hon-shimeji (*Hypsizygus tessulatus*).

The challenge mushroom farmers face is converting wild mushrooms to domestic cultivation. This is more difficult than it may seem. Although white mushrooms will take root in beds of straw and manure, and oysters and shiitakes can be raised on dead tree trunks, other mushrooms, such as boletes (cèpes or porcini), chanterelles and truffles, live in symbiotic relationships with the root systems of living trees — a tricky situation to replicate artificially. Progress has even been made in cultivating fresh morels (*Morchella esculenta*), but there is still a way to go in formulating the ideal growing medium before these wild delicacies can become a supermarket staple.

Whatever the variety, mushrooms do not grow from seeds, but from spores. To grow white mushrooms, these spores are planted in big trays of specially formulated, sterilized compost made from the selected refuse of several different industries: cocoa bean hulls, cottonseeds, straw and ground corncobs. These beds are stacked several deep in long, windowless cinder block buildings. Two to three weeks after the spores are planted, the root systems of the mushrooms take hold. Called mycelia, the roots resemble lacy white threads. The mycelia are covered with peat moss, and within a couple of more weeks, the first mushroom caps begin to emerge. These are allowed to mature for another week or so before harvesting. The mushroom harvest takes place over about a week, and then the "mushroom house" is stripped down and sterilized before a new cycle begins. All told, it takes between a month and a month and a half to grow a mushroom crop from beginning to end.

The size of a mushroom is not dictated by its age. Mushrooms of the same maturity can range from a tiny button to a larger cap. The best indicator of age is the degree to which the mushroom has flattened out, exposing the gills underneath.

The portabello mushroom is simply a brown crimini mushroom that is allowed to mature for another week. This results not only in the mushroom's being bigger (portabellos can weigh as much as a quarter pound each, whereas it takes 25 to 30 crimini to make a pound) but also in completely exposed gills and a softer texture. For years these mushrooms were regarded as expensive mistakes — a sign that a mushroom house had been poorly maintained — and were discarded as not worth selling.

Then, in the early 1990s, a marketer got the bright idea of selling them to restaurants as "domestic wild" mushrooms. The name "portabello" was invented because it sounded Italian (it is a completely made-up word, and there is no agreement on spelling; you'll see them labeled "portobello," "portabella," "portobella" and "portabello"). The idea took off. So popular have portabellos become that crimini are now sometimes marketed as Baby Bellos.

When you taste portabellos and crimini side by side, it's hard to tell whether there is much of a difference in flavor. But there is a big difference in texture: crimini are much firmer, while portabellos are meatier. Crimini are better for using raw, as in salads, while portabellos are better for grilling. There is one other difference: portabellos have fully mature gills, which stain everything they touch a rather dismal gray. The gills are easily cut away, though.

Mushrooms have become popular well beyond anyone in Kennett Square's wildest dreams. Since 1971 American per capita consumption has increased by more than eightfold. Americans now consume on average 2.6 pounds of fresh mushrooms every year. That is enough to make mushroom growing a more than $900-million-a-year business. Mushrooms are the fourth-leading vegetable in terms of value to farmers, behind only potatoes, tomatoes and lettuce. White buttons make up the vast majority of sales (better than 85 percent), but sales of brown mushrooms — both crimini and portabellos — are growing fast, as consumption of brown mushrooms has nearly doubled in the past decade.

WHERE THEY'RE GROWN: More than half of the mushrooms grown in the United States come from Pennsylvania. The next leading producer is California, with about 14 percent.

HOW TO CHOOSE: Mushrooms are predominantly water — between 80 and 90 percent — and the "skin" that covers their exterior is extremely thin. As a result, mushrooms dry out very quickly. Choose mushrooms that are smooth, glossy and wrinkle-free. Also avoid any that have obvious bruising, as damaged flesh breaks down very quickly.

HOW TO STORE: Store mushrooms in a paper bag or some other moisture-absorbing container in the refrigerator. Any moisture that collects on the mushrooms will quickly cause spoilage.

HOW TO PREPARE: For most mushrooms, you need to do little more than wipe them clean and cut away the hardened base of the stem. Contrary to popular belief, you can rinse mushrooms with water without causing any damage. (They are mostly made up of water to begin with.) But to avoid spoilage, do this just before you cook them, then wipe them dry so they'll brown during cooking. Like apples and potatoes, mushrooms are very prone to enzymatic browning. Cut them up at the very last minute. If you must cut them up in advance, toss them with a little lemon juice to delay browning.

ONE SIMPLE DISH: When cooking mushrooms, here's a good trick to get the best flavor: start the mushrooms in a hot pan and add the seasonings only after they've cooked a bit. When the mushrooms hit the hot butter, they'll start to give off moisture. If you add the seasonings at this point, their flavor will carry back to the mushrooms as the liquid concentrates and is reabsorbed. Using a hot pan allows the mushrooms to brown a bit before they start to become limp. Mushrooms prepared this way are terrific as a side dish, served by themselves. A final gloss of butter and a couple of drops of sherry vinegar will round out the flavor nicely. Vary the herbs — maybe some rosemary (just a hint) or tarragon (as much as you want). You can use shallots or garlic. And sometimes it's nice to toss in some chopped toasted hazelnuts for a bit of crunch.

Mushroom, Fennel and Parmesan Salad

This salad pairs soft mushrooms and crisp fennel. Shaving the cheese with a vegetable peeler results in thin shards that look wonderful.

6 SERVINGS

- 1 large fennel bulb (about 1 pound)
- 1 pound button mushrooms
- 4 sprigs thyme
- ½ cup olive oil
- ¼ cup fresh lemon juice
- ¼ cup minced green onions
- ¾ teaspoon salt
- ¼ teaspoon minced garlic
- Freshly ground pepper
- 2 ounces Parmigiano-Reggiano

Cut the stalks and fronds from the fennel and reserve enough fronds to make ¼ cup when chopped. Cut the bulb lengthwise into quarters and trim away the triangle of solid core at the base. Cut each quarter lengthwise into thirds or quarters and cut each of those in half crosswise.

Trim the tough bottoms of the mushroom stems and slice the mushrooms just under ¼ inch thick. Strip the leaves from the thyme between your finger and thumbnail.

In a large bowl, whisk together the olive oil, lemon juice, thyme leaves, green onions, fennel fronds, salt and garlic. Add the fennel and stir to coat. Remove the fennel with a slotted spoon, draining any excess dressing back into the bowl, and arrange the fennel on a platter in a broad layer.

Add the mushrooms to the leftover dressing, season with pepper to taste and stir to coat well. Arrange the mushrooms in an oblong mound on top of the fennel, centering the mound so the fennel shows around the edge.

Using a vegetable peeler, shave the cheese over the salad and serve.

Mushroom Hash

If you're making this with fully opened portabello mushrooms, take a couple of minutes to scrape away the dark gills with a small spoon. That will keep them from staining the mixture with their ink. If the mushrooms are still closed, you needn't bother.

6 SERVINGS

1½ **pounds mixed mushrooms (portabello, crimini, maitake and shiitake)**
3 **tablespoons butter**
   Salt
1 **garlic clove, minced**
1 **tablespoon minced fresh parsley**
½ **cup dry white wine**
6 **sprigs thyme**
2 **tablespoons heavy cream**
1½ **pounds mixed fingerling and small boiling potatoes**
½ **teaspoon sherry vinegar**
   Freshly ground pepper

Wipe the mushrooms clean, trim any hard stems and cut the mushrooms into roughly almond-size pieces. Try to use a mixture of sizes of mushrooms so that you can get a diversity of shapes. The small ones can be left whole, those that are a little bigger can be cut in half, and so on.

Heat 2 tablespoons of the butter in a skillet over medium-high heat until the foam subsides and the butter turns a light hazelnut color. Add the mushrooms, sprinkle with ½ teaspoon salt, cover tightly and cook, tossing occasionally, until the mushrooms begin to glisten and give up their moisture, about 3 minutes. Remove the cover, add the garlic and parsley, raise the heat to high and continue cooking, stirring constantly, until the mushrooms are richly aromatic and soft but not flaccid, about 3 minutes.

Empty the mushrooms into a bowl and add the white wine to the skillet. Cook over high heat until it reduces to a syrup, about 2 minutes. Strip the leaves from the thyme between your finger and thumbnail and add them to the syrup along with the cream. Cook, stirring, until smooth, about 5 minutes. Add the mushrooms to the sauce, toss to coat well and set aside.

Cut the potatoes into ½-inch pieces and steam in a tightly covered pot over rapidly boiling water until just tender, about 15 minutes.

Warm the mushroom mixture over medium heat and add the potatoes as soon as they are done. Do not let them cool, or they won't absorb the flavors. Add the sherry vinegar and stir everything together. Season with salt and pepper to taste. (The dish can be prepared to this point up to 1 hour in advance.)

Warm the mixture over medium-high heat, add the remaining 1 table-spoon butter and stir to mix well. Serve.

Peppers

. . . .

If you wanted to pick out a single starting point for the produce revolution, there is no shortage of candidates: tomatoes, fresh herbs, baby lettuces. But what about peppers? Chile peppers, certainly, but bell peppers even more so. Thirty years ago bell peppers came in one color — green.

But then a funny thing happened: consumers were offered a choice. And they responded. Between 1960 and 1990 American per capita consumption of bell peppers quadrupled. According to the USDA, on any given day almost a quarter of Americans will eat a bell pepper or a dish containing bell peppers. That's nearly double the percentage of those who will eat a French fry and almost the same percentage as those who will eat a tomato.

This increase happened with chile peppers as well, though on a smaller scale. And oddly enough, after an initial rush of growth, chile pepper production and consumption leveled off in the 1990s, while bell peppers just kept on growing.

How did this come about? Once again, credit that most unlikely of all agricultural powers, the Dutch, who figured a way to outsmart nature. All peppers start out green; it is only as they mature that they begin to show their true colors. Technically what happens is not unlike that which causes the turning of the leaves in New England. Green chlorophyll is a dominant pigment and tends to overshadow everything else. But as fruits begin to mature and develop sugar, that sweetness alters their chemical makeup. The chlorophyll starts to break apart, which allows the underlying colors to reveal themselves. This happens easily in nature, but it takes some doing in agriculture. Peppers are tender and prey to all sorts of bugs and viruses. It takes extraordinarily careful farming to grow peppers to full maturity out-of-doors without their suffering some sort of damage, even if it is just cosmetic. In Holland farmers were consistently able to grow their peppers to full maturity inside greenhouses, meaning that they could offer peppers in a rainbow of colors. Red came first, then yellow, then orange, purple and even something called chocolate (mostly marketing poetry; really more of a bruised green).

Those color changes are merely symptomatic of a deeper transformation: what had been a pretty one-dimensional "green" flavor becomes sweeter and more complex. To chemists it's the breakdown of the one-note 2-isobutyl-3-methoxypyrazine. This is a particularly pesky chemical, detectable to humans in very small concentrations and a symptom denoting a flaw in both cheeses and wines. As peppers ripen, a variety of smells take this chemical's place — some of which comes from those once-hidden pigments that have suddenly become visible. All peppers have these shades of flavor, but we usually recognize only two: hot and sweet.

Perhaps we would be wise to leave our description of peppers there. The wilds of pepper taxonomy are ventured into only by the very brave or very foolish. Peppers are among the most diverse plants, with twenty-two wild families and five domesticated ones. Walk into any Mexican market, and you'll find three or four fresh green chiles and at least half a dozen dried red chiles. (Mature red chiles are almost always dried, both for preserving and because they're usually ground into a powder.

Names change by region. The fresh green chile that one group has always called a pasilla is called a poblano by another group (who use the word "pasilla" to refer to a dried red chile). Ironically, almost all of the peppers we find in the United States — from the common green bell to the fiery jalapeño — are members of the same species: *Capsicum annuum* (the main exception being the incandescent habanero, or *C. chinense*).

Chile peppers have a wide range of tastes, from the sharp, grassy flavor of serranos to the rounder, sweeter flavor of jalapeños.

Drying seems to exaggerate the flavor nuances, much as fermenting brings to the fore the subtleties in wine grapes. Some chiles taste frankly of fruit, some of chocolate, others of smoke (although the truly smoky peppers, chipotles, are ripe jalapeños that have been dried over a smoldering fire).

The one flavor all chiles share, of course, is heat. Pepper heat comes from a chemical called capsaicin, an irritant that causes an intense burning sensation when it touches nerves, whether those nerves are in your mouth, on your hands, in your eyes or anywhere else. That is why chile peppers taste so hot. It is also why pepper spray (essentially atomized capsaicin) is such an effective deterrent. When the active ingredient capsaicin is rubbed into your skin, it burns. But then, according to scientists, the receptors that are sensitive to capsaicin get overloaded and become numb.

Capsaicin is found in significant concentrations only in certain parts of the pepper — primarily the placental veins that attach the seeds to the wall of the fruit. Despite the common cook's advice to remove the seeds to reduce the heat of a pepper, capsaicin is barely detectable there — although if you remove the seeds, you usually remove the veins as well.

Different peppers contain different concentrations of capsaicin, ranging from the extremely potent habanero (the hottest in common commercial production) to the bell pepper (which has all the fire of a cucumber). The presence of capsaicin, and to some extent its concentration, depends on a single recessive gene. But it's not as simple as that. Chile pepper heat is devilishly complicated. Not only does it vary within a single variety, but it also varies within a single farm.

And it can even vary within a single plant. Two seemingly identical peppers picked from the same plant at the same time can contain different amounts of capsaicin. Predicting which pepper will be hot and which mild is a problem that has confounded chile farmers for centuries. The differences can be quite extreme. Little Japanese shishito peppers, for example, are mostly sweet and mild. But about one in every dozen will be hot enough to lift off the top of your head. Plant breeders are working on controlling heat, and they have achieved some success. In the 1990s breeders in Texas succeeded in producing a jalapeño with all the bite of a bell — a great boon to people who say they want salsa but really prefer ketchup.

WHERE THEY'RE GROWN: More than two thirds of the bell peppers grown in the United States come from just two states — California and Florida, with the former holding a slight edge. Peppers are also an important import. About 20 percent of the American supply comes from outside the country, mostly from Mexico. But a surprising number of peppers come from Canadian greenhouse growers, whose exotically colored specialty peppers have increased their American market share more than tenfold in the past fifteen years. When it comes to chile peppers, almost half of our fresh consumption comes from Mexico, whose exports to the United States have increased by 82 percent in the past decade, now totaling more than 425 million pounds.

HOW TO CHOOSE: Even though peppers of all colors are delivered to our markets year-round, from all over the world, you need to be careful when choosing them. Here's how to pick a perfect pepper, spicy or not. First, pay attention to the color. Red peppers should be dark, almost brick-colored; the flavor of the lighter ones isn't as deep. Look at the flesh carefully; it should be uniformly firm. Red peppers have to hang on the plants for a long time to get fully ripe. (They take as much as a month longer to grow than green bell peppers, hence the higher prices.) This extra hang time also allows plenty of opportunity for the kinds of nicks and dents that encourage spoilage. Try to pick out the boxiest peppers — the ones with the flattest sides. Those with graceful undulations look sensuous in Edward Weston photographs, but the skin in those little crevices and hollows will be hard to remove. Finally, hold the pepper in your hand. The best peppers — the freshest ones that still retain the most moisture — are those that are the heaviest for their size.

HOW TO STORE: Store peppers tightly wrapped in the refrigerator, but not in the coldest part. The best temperature for avoiding the water loss that is the most common problem for peppers is 45 degrees — almost exactly the temperature of most refrigerators. If kept colder than that, they can suffer breakdown due to excessive chill.

HOW TO PREPARE: Peppers can be sliced raw and mixed into salads or served with dips. But their flavor improves immeasurably with cooking.

ONE SIMPLE DISH: Stuff roasted and peeled bell peppers or mild poblano chiles with a spoonful of soft fresh goat cheese and snipped fresh chives.

Roasted Red Peppers Stuffed with Tuna

This combination sounds odd, but it is a common antipasto in Italy's Piedmont region, where I first had it at Da Guido, for years perhaps the finest traditional Piedmontese restaurant. The combination of flavors is ethereal — like a creamy tuna mousse wrapped in earthy roasted peppers. Matt Kramer, in his book *A Passion for Piedmont*, makes it a little differently, using half butter and half olive oil, leaving out the capers and substituting basil for parsley. That's good, too. This dish requires good tuna, packed in olive oil. Under no circumstances try this with tuna packed in water.

6 SERVINGS

- 1 6-ounce can tuna packed in olive oil
   Olive oil
- 4 teaspoons capers, with their liquid
- 2 teaspoons chopped fresh parsley
- 3 red bell peppers, roasted, peeled and cut in half lengthwise (see page 262)
   Salt

In a food processor, puree the tuna with 1 tablespoon olive oil until it is smooth. Scrape down the sides and continue to process, slowly adding another tablespoon of oil through the feed tube. The mixture should be almost completely smooth, with a light, foamy texture and a pale color. Scrape the mixture into a small bowl and stir in the capers and their liquid and 1 teaspoon of the parsley.

Place a pepper half on each of six small serving plates. Spoon 1 to 2 tablespoons of the tuna mixture in the middle, then fold over the top. Drizzle lightly with olive oil, sprinkle with the remaining 1 teaspoon parsley and sprinkle lightly with salt. Serve at room temperature.

HOW TO ROAST A PEPPER

Roasted peppers are utterly unlike raw ones. In the first place, roasting removes the thin skin of cellulose on the surface. That's the difficult-to-digest part. And roasting gently cooks the meat, softening it and bringing out hidden dimensions. You can roast a pepper using any number of methods. Perhaps the simplest is just to throw it on the grill. This has the advantage of accommodating a large number of peppers at the same time. A regular 21-inch kettle grill will easily hold more than a dozen large peppers at once. Just keep turning them to hit every bit of skin (including the bottoms and the tops) and move them from place to place so every pepper gets its turn over the hottest parts of the fire. You're looking not for browning here but for a definite blackening: go ahead and char them. So tough is that cellulose skin that even after this rough treatment, when you peel it off, there will be uncharred flesh underneath. Roasting peppers on the grill will take from 25 to 35 minutes, depending on the heat (it's a good thing to do while you're waiting for a really hot fire to die down enough to cook meat).

You can roast large batches even more easily in the oven if you're willing to forgo that smoky grace note (indeed, in recipes such as Roasted Red Peppers Stuffed with Tuna (page 261), a purer flavor is better). To roast peppers in the oven, arrange them on a jelly-roll pan and bake them at 400 degrees, turning them once or twice to keep them from sticking. That will take 20 to 30 minutes. Cooked this way, the skin will puff up like a balloon, without blackening near as much as it does when grilled. Then cover the peppers with a damp cloth and let them cool for 10 to 15 minutes. The steam will finish loosening the skin.

Some cooks recommend roasting peppers under the broiler or over an open flame on a stovetop burner. While both of these methods will work, they have significant drawbacks. The broiler cooks the peppers far too quickly and unevenly. The center row under the flame will be done while the next row is still practically raw. Doing them on the stovetop has the obvious disadvantage of letting you do only one or two at a time. And heaven help you if a roasting pepper pops, as they are wont to do, spilling their juices so they bake onto the stove.

Peel the peppers by rubbing away the charred skin with your fingers. For tough spots that might be a little underdone or are in hard-to-reach crevices, use the back of a knife. Though you may be tempted to rinse the peppers to get rid of the last little flecks of skin, don't. The flesh is coated with a thick, delicious juice, and you don't want to lose any of it. Be careful to dump the skins into the trash can; don't put them down the disposal. Few things will clog plumbing faster than pepper skins. The pigment from red peppers is incredibly resilient, too, so it takes a lot of scrubbing to get the stain out.

Salad of Roasted Peppers and Ricotta Salata

Because roasted peppers are so sweet and unctuous, it's a good idea to pair them with spikier ingredients that are sharp or salty, such as olives, capers or anchovies. Ricotta salata is a moist, firm, tangy cheese. If you can't find it, use dabs of soft fresh goat cheese instead.

6 SERVINGS

4	red bell peppers, roasted and peeled (see page 262)
6	green olives, pitted and cut into slivers
2	salted anchovy fillets, minced
1½	teaspoons sherry vinegar or red wine vinegar
1	garlic clove, minced
½	teaspoon salt
1	tablespoon olive oil
1½–2	ounces ricotta salata

Tear the peppers into generous bite-size pieces. (Cutting is too uniform and spoils the appearance.) Toss together the peppers and olives in a medium bowl.

In a small bowl, combine the anchovies, vinegar, garlic and salt and whisk to combine. Drizzle in the olive oil as you continue to whisk the mixture.

Pour the dressing over the peppers and toss to coat well. Turn the salad out onto a serving platter. Using a vegetable peeler, shave long shards of the ricotta salata over the top. Don't be stingy with the cheese. Serve.

Peperonata

This is one approach to the basic stew of tomatoes, onions and peppers found all over the Mediterranean, for which there are as many recipes as cooks. The final fillip of the chile, garlic and herb sauce is inspired by a recipe in Patience Gray's *Honey from a Weed*. Using yellow bell peppers relieves the monochromatic brick color of cooked tomatoes and red peppers, but with or without them, this is one delicious dish. Try serving it with a grilled lamb chop and a glass of young Zinfandel, Chianti, or Dolcetto.

6 SERVINGS

- 5 tablespoons olive oil
- 1 onion, sliced
- ¾ cup chopped tomatoes
- 5 bell peppers (preferably a mixture of red and yellow), roasted and peeled (see page 262)
- ¾ cup dry red wine
- Salt
- ½ jalapeño chile, roasted, peeled and seeded
- 2 garlic cloves
- 1 cup chopped fresh basil leaves
- ¼ cup chopped fresh parsley

Warm 2 tablespoons of the olive oil in a large skillet over medium heat. Add the onion and cook until it softens, about 5 minutes. Add the tomatoes and bell peppers and cook briefly. Add the red wine and ½ teaspoon salt, cover and cook, stirring occasionally, for about 10 minutes. Remove the lid and continue cooking, stirring occasionally, for another 10 to 15 minutes. Check frequently toward the end, as the peppers will want to stick to the bottom of the pan.

While the bell pepper mixture is cooking, pound the jalapeño and garlic to a paste in a mortar with ½ teaspoon salt. Add the basil and parsley and pound to a paste. Add the remaining 3 tablespoons olive oil and stir, grinding more with the pestle. The finished sauce should have a rather loose consistency, somewhat more liquid than pesto.

Stir the sauce into the bell peppers, taste for salt and heat through, about 2 minutes. This is equally good served hot, cold or anywhere in between.

Chile and Zucchini Braised in Cream

Simmer together a poblano chile, long-browned onions and Mexican crema fresca, and you can make almost anything taste heavenly. I love to add steamed new potatoes. Raw corn is also wonderful.

SERVES 6

- 1 onion
- 2 tablespoons vegetable oil
- Salt
- 2 pounds zucchini
- 1 poblano chile
- 2 garlic cloves, thinly sliced
- ½ cup crema fresca, crème fraîche or sour cream
- 4 tablespoons chopped fresh cilantro

Cut the onion in half lengthwise and then cut it crosswise into ¼-inch slices. Place oil and onion in a heavy-bottomed skillet and sprinkle with salt. Cover tightly and cook over very low heat until the onion has thoroughly wilted and begun to turn golden, about 30 minutes.

Meanwhile, cut the ends off the zucchini, quarter it lengthwise and then cut the quarters in half crosswise. Roast the poblano chile (see page 262) until the skin is blackened and blistered. Set aside to cool. When the chile is cool enough to handle, peel, core and seed it, then cut the flesh into shreds.

When the onion is golden, add the garlic and cook for 5 minutes more. Add the poblano and zucchini and stir to combine well. Add ¾ teaspoon salt, cover and continue to cook over low heat until the zucchini begins to soften but is still crisp in the center, about 20 minutes.

When the zucchini is ready, pour the crema fresca over the top and raise the heat to medium-high. Cook, uncovered and stirring occasionally, until the cream has thickened, about 10 minutes. If the cream begins to brown and stick to the bottom of the pan, scrape it free with a spatula; it will dissolve and enrich the sauce.

Just before serving, taste and adjust the salt. Stir in 2 tablespoons of the cilantro. Transfer to a serving bowl and sprinkle with the remaining 2 tablespoons cilantro. Serve.

Winter Squash

. . . .

Winter squash varieties are so different that sometimes it's hard to believe they are related. Their skins may be rough and warty or smooth and sleek, their shapes round or cylindrical. Their colors can range from orange and yellow to green and nearly blue — or just about any combination or variation thereof. Their flesh may be stringy and fibrous or smooth as butter, and their flavor can be sweet and rich or thin and vegetal. Winter squash vary in size from a little bigger than a tennis ball to more than one hundred pounds. Even the name "winter squash" is a misdirection — they are actually at their best in the fall (hence all those Halloween pumpkins). The vegetable acquired its trans-seasonal identity because back in the bad old days before refrigeration (and air shipment), it was one vegetable that could be relied on as a staple late into the frozen months.

Though native to the Americas, squash are grown and loved all over the world — in Europe, of course, but also in Asia and Africa. Partly because of this geographical proliferation, a final count of the many varieties is probably impossible. There are three major species of winter squash: *Cucurbita pepo*, *C. maxima* and *C. moschata*. If you want to tell which is which, check the stems: *pepo* stems are angular and flared where they attach to the squash; *maxima* stems are round; *moschata* stems are smooth and grooved. Each species is broken up into dozens of separate varieties. North Carolina State University's agriculture department lists more than 350 varieties grown in North America alone. Naming is a quagmire. It seems more the rule than the exception for a single type to have a couple of different names. At the same time, some squash names represent several different varieties. To cite just one example, the name "kabocha" is not only generic but also redundant. It simply means "squash" in Japanese, and the family really consists of several closely related varieties, variations of *C. moschata* (usually called "Japanese pumpkin") and *C. maxima*. Some of the most popular types are combinations of the two.

So what exactly is a winter squash? All the varieties have a couple of things in common. Although they are usually eaten as vegetables, they are actually vining fruits (remember, they contain seeds). They are members of the Cucurbit family, which includes summer and winter squash, melons and cucumbers. Unlike summer squash, winter squash are picked at full maturity, after they have developed a hard shell. (It is this key attribute that discourages spoilage and accounts for the long shelf life.) Many winter squash (though not all) actually improve in both sweetness and general flavor with at least a couple of weeks of storage. They all tend to have golden yellow to buttery orange flesh. On a purely subjective basis, they are almost uniformly beautiful, with rich colors and textures that make them look like elaborately shaped pieces of rustic pottery. But beyond those few similarities, pretty much anything goes.

How do you make sense of it all? The good news is that when you get right down to it, it's not so hard to pick a winter squash because, quite frankly, you can ignore most of the varieties. Beautiful as they are visually, most were traditionally prized more for their keeping qualities than for any outstanding culinary characteristics. To my taste, there is no better example of that than the hallowed pumpkin, which, in most cases, is a singularly stringy, watery, vegetal-tasting mess. (The squash variety grown for commercial pumpkin pie filling is actually closer to a butternut squash than anything you might recognize as a pumpkin.) So unless you are a cucurbit completist, you should focus on a handful of the best varieties that are commonly available in your area. Several excellent squash may be offered on a very limited basis, such as Hubbard, red kuri and buttercup. And if someone whose taste you trust recommends another variety, by all means give it a shot. In general, however, I suggest the following four varieties, which are both delicious and easily found. (Winter squash seem to be defined by two variables: *texture* — from stringy to smooth — and *flavor* — from a nutty sweetness to a kind of green vegetal flavor I'll call "squashiness" for lack of a better term.)

- Acorn. Probably the most familiar winter squash after the pumpkin, it's certainly the most familiar delicious one. The acorn is a middle-of-the-road squash. The skin is dark green with occasional blushes of saturated orange, the flesh is pale to medium orange, the texture is semismooth and rich and the flavor is moderately sweet and moderately squashy. Table Queen is an especially good type of acorn.

- Butternut. If I were forced to choose a single readily available winter squash variety for cooking, this would be it. Butternut is shaped like a long cylinder with a slight bulb on one end. You're usually best off choosing one with the fattest neck and the smallest bulb because it will have the smallest seed cavity and the most meat. The skin is fairly thin and golden tan in color, the flesh is dark orange and semifibrous and the flavor is very sweet and nutty, with just a hint of green squashiness.

- Carnival. This one looks like a harlequin acorn squash, with beautiful patchwork dark green and bright orange skin. The flesh is dark orange and slightly fibrous, and the flavor is complex, rich and sweet, with an intriguing earthy note.

- Kabocha. Although kabocha has become widely available only within the past decade or so, it seems to be everywhere today. This squash is round and slightly flattened at both ends. The skin is dark green with delicate gray-blue tracing (there are also all-green and dark orange versions), the flesh is pale to medium orange and extremely dense and smooth and the flavor is very sweet, with a nice green squashy edge for some backbone.

Whatever the variety, picking a particular squash is a bit of an art. One of the best clues is to inspect the stem, which should always be present and should be dry and corky. This tells you that the squash stayed on the vine until it was almost ready to fall off. (Botanists call this natural separation "abscission.") The color of the skin should be deep and vibrant, which shows the full development of the chlorophyll and carotenoid pigments that come with maturity. The quality of the color should be matte rather than shiny. Many squash show yellow or golden spots where they rested on the ground, just as melons do. Although this area may be pale, it should be deeply colored as well, and certainly not green. When a squash is fully mature, you won't be able to nick the skin with your thumbnail.

"Curing" the squash — storing it for a couple of weeks under the proper conditions — improves the flavor of some varieties. During this period enzymes convert much of the squash's starch to sugar. Indeed, one study found that proper curing (at 75 to 80 degrees and high humidity) for up to three weeks had more effect on the sweetness and flavor of some squash than did an extra week on the vine. This is particularly true of *moschata* squash, such as butternut and kabocha, and true to a slightly lesser extent of *maxima* squash, such as red kuri, Hubbard and pumpkins.

Pepo squash, such as acorn, carnival, spaghetti, delicata and sweet dumpling, are closely related to zucchini and other summer squash. They can be cured to harden the shell and reduce the moisture content (improving texture), but they do not convert starch to sugar. They also have thinner skins and do not store as well.

At the market you can spot cured squashes because their colors, though saturated, will be slightly faded. After curing, squash should be stored at cool room temperature (about 50 degrees). Refrigerating them will deaden the flavor and cause pitting and soft spots to occur on the surface.

The flavor and texture of winter squash will vary tremendously depending on how it is cooked. When it is cooked with moisture, such as in steaming, the taste is subtle and the texture delicate. (Normally, you wouldn't want to simmer squash fully immersed unless you're making soup, because the flesh is so delicate it will start to dissolve in the liquid.) When it is cooked with dry heat, such as in roasting, the natural sugars caramelize. As the mois-

ture evaporates, the flesh becomes intensely sweet and deeply flavored, and the texture becomes dense and creamy, even buttery. If
you've never tried sautéing winter squash, you should. The exterior
caramelizes nicely, but the interior stays delicate. Even better, instead of taking 45 to 60 minutes to cook (as with roasting), sautéed
squash is done in less than 15 minutes.

WHERE THEY'RE GROWN: Most winter squash are at their
best from late September through early November. Later than
that, you should stick with butternut and kabocha, which will last
through December and even into early January. Winter squash in
general are not grown widely enough to be tracked separately as a
category; pumpkins are grown primarily in Illinois and California.

HOW TO CHOOSE: In general, choose squash that have a
hard shell; deep, vibrant colors; a hard, corky stem; and a
deeply colored resting spot.

HOW TO STORE: Store winter squash in a cool, dark place.
Do not refrigerate.

HOW TO PREPARE: Getting to the sweet inner meat of a
winter squash can be a challenge. Some varieties have skins
so tough that you have to crack them with a hatchet before you can
begin to cook with them. Obviously, unless you have a hatchet in
your knife block, stick with the tenderer types, such as butternut
or any of the *pepos*. These can be peeled either before or after cooking. Always peel squash before sautéing, however. With some of the
thinner-skinned squash, a reasonably sharp vegetable peeler will
be all that's required. Otherwise, you'll have to use a chef's knife.
After you peel the squash, cut it in half and remove the seeds and
strings. Then dice the flesh neatly. When you're roasting squash,
it's best to remove the peel after cooking. Before roasting, cut the
squash in half and remove the seeds and strings. Place the halves

cut side down in a roasting pan with a little water. Roast at 400 degrees, turning the squash cut side up after 20 minutes. Cook until the flesh is easily pierced with a fork, usually about 1 hour. Once the squash is cooked, spoon the flesh away from the skin, if desired.

ONE SIMPLE DISH: Cut the squash in half and remove the seeds and strings. Place the halves cut side up in a roasting pan and add about ¼ inch water. Place a pat of butter in the cavity of each half, salt lightly, cover tightly with aluminum foil and bake at 400 degrees for 30 minutes. Remove the foil and continue roasting, basting the squash occasionally with butter from the cavity, until the squash is quite tender, about 30 minutes more. Dust lightly with freshly grated nutmeg before serving.

Winter Squash Risotto with Walnuts and Fried Sage Leaves

We usually think of winter squash as having emphatic flavors, but diced fine, it adds a subtle sweetness and earthiness to this risotto. The crisp fried sage leaves perfume the risotto nicely.

6 SERVINGS

⅔	cup walnut halves
	Vegetable oil for frying
18–24	fresh sage leaves
4	cups chicken or vegetable broth
4	cups water
5	tablespoons butter
1	cup finely diced onion
½	pound peeled butternut squash, cut into ½-inch cubes
⅔	cup dry white wine
2	cups Arborio rice
2	teaspoons salt
¼	cup freshly grated Parmigiano-Reggiano, plus more for passing

Heat the oven to 400 degrees. Toast the walnut halves in a dry pan in the oven, tossing occasionally, until lightly browned and nutty-smelling, about 20 minutes. Chop coarsely and set aside.

Pour the oil into a small saucepan to a depth of about 1 inch and place over high heat. When the oil reaches 375 degrees, add the sage leaves and fry just until they darken slightly and turn crisp, only a couple of seconds. Rescue the leaves with a slotted spoon and drain on paper towels.

Combine the chicken broth and water in a large saucepan and bring to a boil. Reduce the heat to maintain a bare simmer.

Place 3 tablespoons of the butter and the onion in a large skillet over medium-high heat. Cook, stirring, until the onion begins to soften, about

3 minutes. Add the squash and cook until it is shiny and beginning to soften, about 5 minutes. Add the white wine and cook until it reduces to a syrup, about 3 minutes.

Add the rice to the pan and cook, stirring, until the mixture is dry enough that the rice makes a "singing" sound as it scrapes the bottom. Add 1 cup of the simmering stock and the salt to the pan and stir it in. Cook until the rice absorbs enough liquid so that you can see the dry pan when you stir, about 5 minutes. Keep cooking this way, adding more stock as needed and stirring, until the rice is firm but tender, with no chalky center. This will take about 20 minutes in all. You don't need to stir continuously, just when you add the stock to the pan and when it is nearly dry.

When the rice is done, add about ½ cup more stock to loosen the mixture slightly. Remove the pan from the heat. Stir in the remaining 2 tablespoons butter and the grated cheese. Stir vigorously to free as much starch as possible, which in combination with the cheese will thicken the risotto slightly.

Spoon the risotto in generous mounds in the center of six wide bowls. Scatter the toasted walnuts over the top. Place 3 or 4 fried sage leaves on top of each bowl. Serve immediately, passing more cheese at the table.

Mushroom and Spaghetti Squash Gratin with Parmesan Bread Crumbs

Okay, I know what you're thinking: how good can something with spaghetti squash be? Miracle of miracles, this gratin is absolutely delicious. Somehow the usually bland vegetal flavor of the squash amplifies the earthiness of the mushrooms in a totally unexpected way. This is very loosely based on a recipe by my friend Deborah Madison.

8 SERVINGS

- 1 3½- to 4-pound spaghetti squash
- 1¼ pounds mixed mushrooms (portabello, crimini and button)
- About 4½ tablespoons butter
- 2¾ teaspoons salt
- 2 tablespoons minced shallots
- 3 slices prosciutto, cut into thin slivers
- 2 leeks
- 1 cup crème fraîche
- 1 cup heavy cream
- ½ round loaf day-old sourdough
- ¼ cup freshly grated Parmigiano-Reggiano (about 1 ounce)

Heat the oven to 400 degrees. Cut the squash in half and remove the seeds. Place the squash cut side down in a roasting pan and add about ½ inch water. Bake until the squash is easily pierced with a knife, about 1 hour. Do not turn off the oven.

Wipe the mushrooms clean, trim any hard stems and cut them into thick slices. Heat 3 tablespoons of the butter in a large skillet over medium-high heat until the foam subsides and the butter turns a light hazelnut color. Add the mushrooms and sprinkle with ¾ teaspoon of the salt. Cover tightly and cook, tossing occasionally, until the mushrooms begin to glisten and give up their moisture, about 3 minutes. Remove the cover and add the shallots. Raise the heat to high and continue cooking, stirring constantly, until the mushrooms are richly aromatic and soft but not flaccid, about 3 minutes. Transfer the mushrooms to a bowl and set aside.

Reduce the heat to low and, without wiping out the pan, add the prosciutto. Cut away the dark green leaves of the leeks, then quarter them lengthwise, leaving them attached at the roots. Rinse thoroughly under cold running water and slice thinly crosswise. Add the leeks to the prosciutto and cover tightly. Let the prosciutto and leeks sweat slowly, stirring occasionally, until the leeks are quite tender, about 10 minutes.

Add the mushrooms back to the pan along with any liquid that has accumulated in the bottom of the bowl. Stir to combine with the prosciutto and leeks. Add the crème fraîche and continue to cook slowly, stirring occasionally, while you clean the squash.

Remove any scorched spots from the cut side of the squash. Hold 1 squash half over a large bowl and, using a fork, scrape out the strands, separating them as you work from one end of the squash to the other. When there is little left but the skin, empty the squash strands into the bowl. Repeat with the other half, adding the squash to the bowl. Season the squash with the remaining 2 teaspoons salt and stir well to combine.

Add the mushroom mixture to the squash and stir to combine. Transfer the mixture to a large gratin or baking dish, mounding it slightly in the center. Add the heavy cream, shaking the pan gently to distribute it through the squash. The cream should just be visible around the edge of the squash. Bake until the cream is bubbling and beginning to darken around the edge, about 15 minutes.

While the gratin is baking, prepare the bread crumbs. Cut away the crusts of the sourdough and cut the interior into cubes. Process in a blender to make coarse crumbs; you should have about 2½ cups. Add the cheese and pulse 3 or 4 times to thoroughly combine.

Scatter the bread crumbs evenly over the gratin, then dot with about 1½ tablespoons butter. Return to the oven and bake until the top is golden brown, about 15 minutes. Let cool slightly before serving.

Caramelized Winter Squash with Rosemary Gremolata

Serving winter squash doesn't mean you have to start preparing it hours in advance. This recipe is done in about 20 minutes from start to finish. The minced rosemary, garlic and lemon need to be cooked only long enough to release their wintry perfumes.

6 SERVINGS

- 2 tablespoons pine nuts
- 2 teaspoons minced fresh rosemary
- 2 teaspoons minced lemon zest
- 1½ teaspoons minced garlic
- 2 tablespoons fresh lemon juice
- 3 tablespoons olive oil
- 2 pounds peeled winter squash, cut into roughly ½-inch cubes
- Salt
- Freshly ground pepper

Toast the pine nuts in a small skillet over medium heat, stirring, until they are lightly browned and fragrant, about 10 minutes. Set aside.

Combine the rosemary, lemon zest and garlic in a small bowl and add just enough lemon juice to moisten. Stir together roughly with a spoon, crushing and smearing them to make a thick paste. The pieces of garlic and rosemary should be extremely fine because they will need to cook in a flash. Set aside the herb paste and the remaining lemon juice.

Heat a large nonstick skillet over medium-high heat. Add the oil and when it is very hot, add the winter squash. Sprinkle with 1 teaspoon salt and toss so that the squash is evenly coated with oil and seasoned with salt.

Cover tightly and cook, without stirring, for 2 minutes. Remove the lid and toss to stir the squash. The cooked sides should be starting to caramelize. Cover again and cook for another 2 minutes.

Remove the lid and toss to stir. Reduce the heat to medium and continue cooking, stirring occasionally, until the squash is just tender enough to be pierced with a sharp knife, about 5 minutes. The cubes should appear somewhat glazed and browned on most of their surfaces, but they should not be so well cooked that they are falling apart.

Sprinkle the herb mixture and the remaining lemon juice over the squash. Toss to coat, letting the herb mixture sizzle briefly and become aromatic. Taste for salt and lemon juice. Season with pepper to taste. Scatter the toasted pine nuts over the top and transfer to a serving bowl. Serve.

Apples

. . . .

Americans have always been proprietary about apples.
More than any other fruit, they serve as an icon of many of the
characteristics we hold most dear: hardiness, pioneering spirit and
just plain goodness. But sometimes when you identify with a thing
too much, you take it for granted. And that's just what happened
with apples.

For most of the twentieth century, the United States dominated
the world's apple trade. We shipped apples around the globe, making Red and Golden Delicious household names from Argentina to
Italy to Taiwan. Today the United States has lost that lead, and by
an overwhelming margin. China, a country that barely grew any
apples to speak of thirty years ago, now grows one third of all the
apples in the world — more than five times as many apples are
harvested in China every year as in the United States. American
apple exports fell by a third between 1994 and 2004, dropping the
United States into third place behind China and Chile.

And those juggernaut American apple varieties Red and Golden
Delicious? They're fading, too. Granted, they are still the most common apples in the United States, but they no longer tower over all
the others as they once did. Instead, they are being shoved aside by
new, often imported varieties. Red Delicious, which once made up
almost half of the national harvest, now accounts for only a little
more than a quarter. Instead, the market is crowded with new and
perhaps unfamiliar types such as Gala, Fuji, Honeycrisp and Pink
Lady.

How did all this come about? Like anything involving world trade and agriculture, the answers are many and complicated. But if you need a one-word summary, "complacency" would be a good place to start. The recent history of the American apple industry is a good example of what can happen when you think you have the world on a string.

At one time hundreds of apple varieties were grown in the United States. Each region had its own stable from which to choose. But the modern American apple industry, which has been dominated by Washington State farmers since the 1920s, was built on two varieties: Golden Delicious and Red Delicious. The former is a very good apple — when it is allowed to mature to the point that it is truly golden and not green. Unfortunately, it usually isn't.

The same cannot be said for the latter, which is about as close to a perfect example of the ruination of industrial agriculture as you will ever find. Originally, the Red Delicious was a pretty good apple. But it was not very red — more golden with red stripes. As is so often repeated in the fruit industry, red sells, so farmers began pushing for Red Delicious varieties with deeper and deeper color — to the point that many of the strains that are grown today are nearly black. What the farmers failed to notice is that as the skin of an apple darkens, it also develops a more bitter taste. Furthermore, when you begin selecting fruit strictly on the basis of color, other attributes fall quickly by the wayside. As a result, the modern Red Delicious is frequently mealy and insipid, with a bitter finish.

Still, it sold and sold. And farmers planted more and more. And apple lovers bemoaned the sorry state of the Red Delicious and the lack of all of those beloved antique varieties they had so adored. But their complaints fell on deaf ears. Growers were selling about as many apples as they could grow — and for pretty good prices — and they weren't about to do anything that might upset the cart. In fact, the commercial quality standards on which wholesale apple prices are based became heavily skewed toward the Red Delicious, rewarding the fruit that was the biggest and reddest with the highest prices.

A closer reading of the situation would have revealed some troubling trends. For instance, apples were not attracting new customers at home: America's per capita apple consumption remained almost flat for the second half of the twentieth century. As a result, more and more fruit had to be sold overseas, in Canada, Mexico and Taiwan. In the 1970s America exported about 6 percent of its apple crop. In the 1980s that increased to 12 percent, and by the turn of the new century exports had increased to 20 percent. In an effort to appeal to overseas buyers, particularly the Taiwanese, American farmers started planting some new varieties that had been developed in the Pacific: Gala (from New Zealand) and Fuji (from Japan).

And then the apple world's sleeping giant woke up. In the early 1980s the Chinese government decided to begin allowing trade with the outside world. Perhaps due in part to the success of Americans selling in Taiwan, apples, which to that point had been a fairly minor crop in China, became a focal point. Apple production in China increased by more than 750 percent from 1980 to 2000. Apples — primarily Fuji, Jonagold, Golden Delicious and Gala — are now the most widely planted tree fruit in China.

As these low-priced Chinese apples started hitting what had traditionally been American export markets, the effect on U.S. farmers was disastrous. By some estimates, as many as half of the apple growers in the United States went out of business in the last twenty years of the twentieth century. In just the five-year span from 1997 to 2002, well after the worst of the crash, the number of farms growing apples declined by 20 percent, and more than 40,000 acres of orchards were pulled up.

By one agricultural economist's calculations, Red Delicious growers lost money on every case of apples sold throughout the 1990s. Another figured that fewer than 15 percent of Washington State apple farmers were earning enough from their harvests to support their families without outside income.

Having weathered the worst of this, American apple farmers are now cautiously optimistic. Because of low quality, poor handling and lack of storage facilities, the Chinese have not been as successful at exporting as it was feared they would be. More important, at least for apple eaters, when those new Pacific varieties of apples could no longer be sold as profitably overseas, they started turning up on domestic shelves. And guess what? It turned out they had pretty good flavor. From being almost unknown twenty years ago, Gala and Fuji are now the third and fourth most widely grown apples in the United States. The Gala harvest has increased by almost 500 percent since the 1990s, and the Fuji harvest has tripled. And that turned out to be just the beginning. Hard on their heels came Jonagold and Braeburn, then Honeycrisp, Empire and Pink Lady (also called Cripps Pink).

Since apples are one of the few fruits that are still sold labeled by variety, American shoppers are having to learn a whole new set of names, including those for old varieties that haven't been heard of in decades. Now it's a rare market in any area cold enough to grow apples that doesn't have at least one farmer selling heirlooms such as Arkansas Black and Cox's Orange Pippin out of a bushel in the back of his truck. But the odds are that the same guy will have boxes full of Fujis and Galas, too. Here's a quick rundown of some of these apple varieties.

• Fuji. A Japanese-bred cross between Red Delicious and Ralls Janet. Introduced in the United States in the 1980s, this big, sweet, crisp apple varies in color from golden to a slight pink blush in cold climates. It holds its shape in cooking and is a great sauce apple with a buttery flavor. It stores well (in fact, some say it improves with storage) and can be good into late summer.

• Gala. From New Zealand, a cross between Cox's Orange Pippin and Golden Delicious, Gala was first introduced in 1965 but began to become popular in the United States in the 1980s. This aromatic, tart apple is golden with pinkish orange stripes. It holds its shape in cooking and is a great sauce apple with a buttery flavor overlaid with a bit of spice. It is among the first apples harvested, usually starting in the middle of August, but it does not keep very well. Don't buy Galas after early spring.

• Braeburn. One of the first of the Southern Hemisphere apples to become popular, the Braeburn was introduced in the early 1950s. Its roots are uncertain; it sprouted up as a chance seedling, probably as a result of cross-pollination between Lady Hamilton and Granny Smith. It's a spicy apple with a nice tart bite and a juicy, crisp texture. It holds its shape in cooking. Braeburns store fairly well and can be good into early summer.

- Jonagold. A New York apple, Jonagold is the offspring of Golden Delicious and Jonathan and was introduced in the late 1960s. It is a tangy apple that can be slightly soft out of hand and cooks to a creamy texture. It's not for storing; buy Jonagolds before spring.

- Empire. This New York–bred apple was introduced in the mid-1960s and has stayed close to home. It is grown almost entirely on the East Coast, and New York State accounts for more than half the harvest. Empire is a cross between McIntosh and Red Delicious. It is a heavy bearer that grows on healthy trees — the kind of cash crop farmers like. The flavor is good, and the apple holds its shape in cooking. It stores moderately well, but buy before the end of spring.

- Pink Lady. Developed in western Australia in the 1970s under the name Cripps Pink, it was dubbed Pink Lady when introduced in the United States in the mid-1980s. Pink Lady is true to name, the skin has a delicate pinkish cast. At its best, this is a superlative apple, crisp and honeyed, with an almost champagne tartness underneath. But its quality is variable — it needs a cold snap to develop full flavor and color. Pink Lady is one of the last apples harvested, usually picked starting in late September. It stores moderately well and can be good until late spring.

- Cameo. An apple with a lot of promise, Cameo was introduced in 1987. It was discovered as a chance mutation in a Red Delicious orchard in Washington State. It is a bright red–striped apple with characteristic white spots. Cameo has good flavor — sweet and mildly tart — but it is most notable for its crispness and staying power. Cameo will hold its crispness under storage for an extended period, even into midsummer. Because of its dense flesh, Cameo takes longer to cook than most apples.

- Honeycrisp. Another hot new apple that is popular with growers, although it hasn't yet begun to show up in supermarkets in any significant numbers. Red with a golden background, it is crisp and sweet and holds its shape in cooking. Honeycrisp was developed at the University of Minnesota and was introduced in

1991. Its harvest is still centered in the northern Midwest. Like Cameo, it is touted partly for its remarkable storage characteristics — its developers claim that it can last up to six months in cold storage, with no special treatment. Curiously, despite having been bred under controlled conditions, Honeycrisp's parentage is a bit of a mystery. Originally, it was thought to be a cross between Macoun and Honeygold, but recent DNA testing has cast doubt on that.

WHERE THEY'RE GROWN: Apples are widely grown, but more than half of the total national harvest and between 60 and 75 percent of the fresh harvest comes from Washington State. New York, Michigan, California and Pennsylvania are also important producers. In Washington, apples are grown in the eastern foothills of the Cascade Range, an apple-producing area since the early 1800s.

HOW TO CHOOSE: All apples should be smooth-skinned and deeply colored. Yellow apples should be golden, and striped apples should have a background color that is nearly golden. Apples should be heavy for their size and firm to the touch.

HOW TO STORE: Apples should be kept as close to 32 degrees and with as much humidity as possible. Store them in an open or perforated plastic bag to retain moisture without collecting water. Put the bag in the crisper drawer of the refrigerator.

HOW TO PREPARE: Apples are often waxed to slow moisture loss and extend their shelf life. This wax is not petroleum based and is harmless, but if you want to remove it, rinse the apple in warm water. Apples can be peeled or not, depending on the dish. Leaving the peel on adds color but can result in stray bits of tough skin in the dish. Core apples before cooking to remove the woody center and hard little seeds; an inexpensive apple corer is the easiest way to do this. Once their flesh has been exposed to air by peeling or slicing, most apples will begin to turn brown. This is harmless in the short run, but it can affect flavor over time. To avoid discoloration, place the sliced fruit in water mixed with a squirt of lemon juice. Red Delicious apples are particularly susceptible to browning.

ONE SIMPLE DISH: Baked apples are one of the homiest desserts. Core the apples, but leave them unpeeled. Stuff the center with a little brown sugar and a knob of butter. Arrange the apples in a buttered baking dish with a bit more butter scattered over the top and pour in a hit of bourbon. Cover with aluminum foil and bake at 350 degrees for about 45 minutes. Remove the foil and increase the heat to 450 degrees. Continue baking until the apples can be pierced easily with a knife, about 15 minutes. If you like, scatter some pecans in the pan during the second part of the baking.

Applesauce with Bourbon, Sour Cherries and Hazelnuts

Can you imagine a better companion to a grilled pork chop than this applesauce? It's also good with beef — either a roast or grilled steaks.

4 SERVINGS

¼ cup dried sour cherries

¼ cup plus 1 tablespoon bourbon

4 apples (about 1 pound), peeled, cored and diced

¼ cup water

1 tablespoon butter

3 tablespoons chopped hazelnuts, toasted

In a small bowl, cover the cherries with the bourbon and set aside to plump while you prepare the rest of the dish.

Place the diced apples and water in a medium saucepan over medium-low heat. Cover and cook until the apples are tender enough to smash with the back of a spoon, 15 to 20 minutes.

Add the butter. Using a wooden spoon or heavy whisk, beat just enough to smash the apples into a thick, chunky sauce. Stir in the cherries and bourbon and continue cooking until the raw alcohol smell of the bourbon has burned off, about 5 minutes. Stir in the hazelnuts and serve.

Gratin of Apples and Dried Cranberries

Apples and cranberries bake in lightly sweetened cream so long that the dish turns into something almost like a caramelized pudding. Meanwhile, the walnuts on top toast and provide a crunchy contrast.

6 SERVINGS

- ½ **cup dried cranberries**
- ¼ **cup orange liqueur**
- 1¾ **cups heavy cream**
- 3 **tablespoons dark brown sugar**
- 8 **cloves**
- 2 **sticks cinnamon**
- 4 **cooking apples (about 1½ pounds)**
- **Juice of ½ lemon**
- 1 **cup walnut pieces**

Heat the oven to 350 degrees and generously butter a 9-inch square gratin dish. Place the cranberries in a small bowl and cover with the liqueur. Set aside to soften, about 10 minutes.

Combine the cream, brown sugar, cloves and cinnamon in a small saucepan and warm over medium-low heat until bubbles appear around the edge, about 5 minutes. Whisk to incorporate the brown sugar thoroughly, then set aside to steep.

Peel and core the apples and slice them ¼ inch thick, then place them in a bowl filled with water and the lemon juice to prevent browning.

Drain the apples and add the cranberries and liqueur to the bowl with the apples. Toss to combine well. Turn the apples and cranberries out into the gratin dish and pour the cream mixture over them through a strainer, discarding the spices. Scatter the walnuts on top and bake until the cream has bubbled and thickened and the apples have browned, about 1¼ hours. Serve warm.

Pears, Asian
Pears and Quinces

Most of the fruits we eat today are modern varieties, genetically selected for commercial agricultural qualities such as shipability, size, color and easy growing. Usually, anything developed even fifty years ago is relegated to the status of "heirloom," as fleeting and scarce as a Duncan Phyfe highboy. That's not the case with pears. In fact, almost every pear variety you've ever eaten is a genuine antique, dating back more than a hundred years. Although there has been some fine-tuning, the household names in peardom have been the same since the middle of the nineteenth century.

The father of the modern pear was a Belgian named Nicolas Hardenpont, who began his work around 1730. Before him, nearly all pears had crisp flesh, more like that of Asian pears than the buttery, melting flesh we so appreciate today. The breeding of pears, like that of apples, is accomplished largely by the propagation of sports — chance genetic mutations that are then refined by horticulturists. So enthusiastic were the Belgians about their pear breeding that, according to one fruit historian, they "seemed to have been quite carried off their feet by [it], and during the first half of the nineteenth century, a fad like the 'tulip craze' of Holland reigned in the country."

The Bosc, that long-necked Gwyneth Paltrow of pears, was developed by Hardenpont's successor, Jean Baptiste Van Mons, in 1807. The commercial workhorse, the Bartlett, came from England, where it was found in a Berkshire church garden in 1770. In London the trees were sold by an orchardist named Williams, who named the variety after himself. (The Bartlett is still called the Williams pear in Europe.) When it came to America, it was renamed by Enoch Bartlett, who began selling it in this country in 1817.

The great American pear, the Seckel, probably originated strictly by chance. It seems that every fall around the beginning of the eighteenth century, a Philadelphia-area hunter named Dutch Jacob used to return from his rounds with the most delicious pears, the source of which he never divulged. He eventually bought the land on which they grew but soon sold it to a man named Seckel, who introduced the pear to the public. The original tree stood until at least 1870. Since it is so different from any native American fruit, it's thought to be a genetic mutation that sprang from the seed of a Rousselet de Reims pear brought over by German settlers.

By the mid-nineteenth century, the roster of pears was fairly set. An orchardist's list from 1857 is full of familiar names: Bartlett, Bosc, Anjou, Forelle, Seckel and Nellis. The only one missing is Comice, which had yet to cross the Atlantic. The queen of pears was discovered by a gentleman farmer in France's Loire Valley, where the inquisitive pear lover can still find a plaque reading: "In this garden was raised in 1849–50 the celebrated pear Doyenne du Comice by the gardener Dhomme and by Millet de la Turtaudiere, President of the Comice Horticole." In his 1934 book *The Anatomy of Dessert,* the English writer Edward Bunyard is palpably overcome by the Comice's grandeur: "In the long history of the pear the year of 1849 stands alone in importance. The historian will be reminded of the annexation of the Punjab, the accession of Francis Joseph, while in that year America hailed her twelfth President in the person of Zachary Taylor. But what are such things to us? . . . Happy those who were present when Doyenne du Comice first gave up its luscious juice to man. Whom could they envy at that moment? Certainly not Zachary Taylor."

As in France and England, there was an explosion of interest in pears and pear growing among the New England landed gentry from roughly 1820 to 1870. Even though the pear is most unsuited to cultivation on the East Coast — being subject to a devastating blight that thrives in the wet, warm eastern summers — it was widely planted. The results were predictable. "It is folly to suppose that every person who plants an orchard of pear trees succeeds," wrote a disheartened P. T. Quinn in his *Pear Culture for Profit,* published in 1869, at the tail end of the craze. "On the contrary, as far as my personal observation has extended, there has been more money lost than made, for I could enumerate five persons who have utterly failed to every one who has made pear culture profitable. . . . It is during the time spent in wading in the dark, without any beacon to guide their steps, that the inexperienced suffer from a series of disappointments."

On the West Coast, though, pears did very well. They were originally brought to the region by Franciscan missionaries (descendants of these original trees were still growing in the orchard at the San Gabriel Mission at the turn of the twentieth century), but modern West Coast pear growing really began with the pioneers who settled Oregon's Willamette Valley in 1847 and the California prospectors who came looking for gold in 1849. The pear thrived in the drier summers there, and with the advent of rail shipment of fruit in 1869, its cultivation became an industry. The West Coast has become the natural home of the Comice, by nature an extremely temperamental fruit that is difficult to cultivate on the East Coast or even in Europe.

"Who does not know the melting Comice, now available so large a part of the year, thanks to the Panama Canal and our own Dominions?" Bunyard asks. "Two thousand years of pear history was necessary to educate a public worthy of such refined delight, and the world's great gourmets had died still unacquainted with the perfect pear."

Why have pears resisted the willy-nilly vagaries of fashion that have so afflicted other fruits? Its survival is due to a combination of factors. First of all, the pear varieties we have are uncommonly good and cover a wide range of flavors and textures. Also, although pears are every bit as easy to breed as, say, apples, they are harder to grow. Pears are susceptible to a wide variety of diseases, so any new variety that is tried has to be proved fit. Finally, introducing a new variety into the marketplace is extremely expensive, and the pear market, as opposed to its seasonal cohort the apple, is relatively small. In fact, just in the past decade, there were a few attempts to "improve" pear varieties, mainly by breeding for red color — something that is largely missing in pears. But those attempts met with a big "ho hum" from consumers, and that has discouraged others from experimenting.

If you want a hint of what pears looked like before Hardenpont tamed them, consider two of the pear's close relatives, the Asian pear and the quince. These are very different fruits, but both have a crisp, granular texture that some describe as "sandy." This texture comes from the distinctive structure of these fruits, whose cell walls are thick and are high in a woody compound called lignin. Pears, if left to ripen fully on the tree, would be hard and woody, too, as their cells develop more lignin as they mature.

As many a home gardener has learned, pears have to ripen after they've been picked to be good. On the tree they ripen from the inside out, meaning that by the time the outside is soft enough to eat, the inside is mushy, almost spoiled. For these reasons, pears are always picked hard. They are then "cured" by chilling to around 30 degrees (because of their sugar content, they won't freeze solid). They are held at this temperature for anywhere from a couple of days (for Bartlett, or summer, pears) to a couple of weeks or more (for Comice, Bosc, Anjou and other pears). After this deep chilling, they are delivered to retail, where they will be ripe and ready to eat in four to ten days, depending on variety and ripening conditions.

A quince is a most unimposing fruit, looking like a blocky, imperfectly carved pear. Most varieties are inedible raw, since they are extremely high not only in lignin but also in tannins, which give the fruit a puckery astringency. What's more, because the flesh is so high in chemical compounds called phenols, it browns extremely quickly once exposed to air by peeling or cutting. Although a few varieties can be eaten out of hand (mostly those, like the pineapple quince, that originated in warmer climates), the common quince must be thoroughly cooked before it can be eaten. A gentle poaching accomplishes one of the more remarkable transformations in the food world. Cooking not only softens the quince's rocky flesh, but it also turns it a rosy pink. Once cooked, a quince is not only beautiful but also delicious, with a warm, spicy perfume.

Asian pears fall somewhere between quinces and European

pears. Although the texture is crisp, not melting, it falls short of being granular. And the Asian pear can be eaten raw and straight from the tree. The flavor is pure pear — honeyed and slightly spicy. To correct a common misimpression, Asian pears are not the result of a cross between a pear and an apple, even though that is a fair description of the appearance of most varieties. Most Asian pears are shaped like slightly flattened apples, though their skins tend much more to bronze and russet rather than shiny red. One popular variety, the Ya Li, looks quite like a Bartlett.

Pears

• • • • •

WHERE THEY'RE GROWN: Pear trees need more cold than most other fruit trees, and they are susceptible to a wide variety of climatic ills. For this reason, between 90 and 95 percent of the total U.S. crop is grown in California, Washington and Oregon. The harvest begins in California's northern San Joaquin Valley in late July or early August. These are Bartlett, or summer, pears, and although they are the first fruit on the market, they are rarely the best. In mid-August to early September, better Bartletts from the cooler Lake and Mendocino counties begin to arrive. Starting in mid-September, you find other varieties of pears grown in the Pacific Northwest. These stay in the market into the following spring but are best before Christmas.

HOW TO CHOOSE: Unlike, say, peaches, the various varieties of pears are very different from one another. Bartletts, the most common variety, have a buttery texture and a mild, sweet flavor. Anjous are firmer and spicier. Seckels are tiny, with a rich taste. Boscs have russet skin and a graceful, slender neck; their flavor is mildly spicy. The Comice is the wide-bottomed queen of the pear family, with a heavenly floral fragrance, a buttery, slightly granular texture and a flavor that is almost winey in its complexity. A pear is perfectly ripe and ready to eat when it is just beginning to soften on the neck, just below the stem. Except for Bartlett pears, pay no attention to color — it changes only very slightly, if at all, during ripening. A perfectly ripe Comice pear, one of the true glories of the fruit world, will still show plenty of green. Bartletts will go from green to golden. They may look scuffed because of their delicate skin, but pay that no mind.

HOW TO STORE: Don't worry if the pears you buy in the grocery store aren't as ripe as you'd like. These are among the best fruits for ripening at home. Just leave them at room temperature until they begin to soften. The process can be speeded up by keeping them in a paper bag to trap the ethylene gas they naturally produce. This also promotes more even ripening. Once they're at the point you like, store them in the refrigerator, loosely wrapped in a plastic bag.

HOW TO PREPARE: The skin of most pears is delicate, so whether to peel them or not is strictly up to the consumer. Pears do have a tough center core that is long and thin, so that usually needs to be removed. And remember that any cut surfaces of a pear will have to be rubbed with lemon juice right away to prevent enzymatic browning, which begins almost immediately.

ONE SIMPLE DISH: If you have great pears, say perfectly ripe Comices, the best thing you can do is put them on a plate with some cracked freshly harvested walnuts and a great blue cheese. Port, anyone?

Asian Pears

• • • • • • • • • •

WHERE THEY'RE GROWN: Almost all of the Asian pears in the United States come from California, although there are scattered plantings elsewhere around the country. Harvest begins in mid- to late July and continues through September. Some varieties, particularly Okusankichi, Shinseiki, Niitaka, Ya Li, Tsu Li, Dasui Li and Shin Li, do well in cold storage.

HOW TO CHOOSE: Even though Asian pears look (and feel) as hard as rocks, they do bruise easily. To prevent this, they are frequently sold with each fruit wrapped in its own little plastic foam cup or sock. Asian pears are picked fully ripe. Russet varieties should be golden brown. Smooth-skinned fruit should be yellow, not green. The Ya Li and other pear-shaped varieties should be pale green.

HOW TO STORE: Refrigerate.

HOW TO PREPARE: Asian pears rarely need peeling. If you do peel them, rub them with lemon juice or keep them in acidulated water to prevent browning.

ONE SIMPLE DISH: Asian pears are best eaten out of hand to fully appreciate their honeyed flavor.

Quinces

• • • • • • •

WHERE THEY'RE GROWN: Less than a thousand tons of quinces are harvested commercially in the United States each year, almost entirely in California.

HOW TO CHOOSE: Fully ripe quinces have a pale golden color and may be covered with a pale gray fuzz.

HOW TO STORE: Store quinces at room temperature. In fact, the more aromatic varieties can be kept in a closet, where they will provide a fragrant perfume until you are ready to cook them.

HOW TO PREPARE: Most quinces must be cooked before they can be eaten. If they are going to be visible in a dish — as opposed to being used in a puree — peel and core them. If you are making a puree or a jam or jelly, leave the peel and core intact, as they are full of pectin, which will help set the mixture. Quinces discolor quickly when peeled or cut, so rub them with lemon juice or store them in acidulated water until ready to use.

ONE SIMPLE DISH: Probably the simplest thing to do with quinces is to poach slices in a simple syrup made of equal parts sugar and water. When they are tender, aromatic and rosy, serve them with plain sugar cookies.

Arugula, Pear and Goat Cheese Salad

I am by nature skeptical of using fruit in savory salads. But this very simple combination has elegance.

4 SERVINGS

Dressing
- 1 tablespoon minced shallots
- 3 tablespoons red wine vinegar
- ½ teaspoon salt
- ½ cup walnut oil

Salad
- ¼ cup walnut pieces
- 2 ripe but firm pears
- ¼ pound arugula or watercress, torn into bite-size pieces
- 6 tablespoons soft fresh goat cheese

For the dressing: Combine the shallots, red wine vinegar and salt in a small, tightly covered jar and set aside to steep until almost ready to use. Just before serving the salad, add the walnut oil to the dressing, fasten the lid tightly and shake well to combine.

For the salad: Place the walnut pieces in a small, dry pan over medium heat. Toast, shaking the pan from time to time to keep the nuts from scorching, until slightly darkened and aromatic, about 10 minutes. Chop coarsely and set aside.

Quarter each pear lengthwise and trim the fibrous core from stem to bottom, cutting out the seeds as well. Cut each quarter in half lengthwise and then cut into roughly ½-inch crosswise pieces.

Place the arugula in a salad bowl and add the goat cheese in rough tablespoon dollops. Add the pears and walnuts and pour ½ cup of the dressing over the top. Mix well but gently, taking care not to smear the goat cheese or break up the pear pieces. Taste and add more dressing, if necessary.

Divide among four chilled plates, being sure each gets an equal amount of cheese, pears and walnuts. Serve.

Pear Clafouti with Pistachio Topping

The texture of a good clafouti is somewhere between a pancake and a custard. It should be moist and eggy, but with a noticeable cakey spring to it. Use this basic recipe, which quite literally whizzes together in seconds in a blender, to complement all kinds of fruit, depending on the season.

6 SERVINGS

¼ cup sugar, plus 1–2 tablespoons for sprinkling
3 large eggs
¾ cup heavy cream
¾ cup milk
½ teaspoon vanilla extract
½ cup all-purpose flour
3 Bartlett pears
½ cup raw pistachios, chopped

Heat the oven to 400 degrees and butter a 9-inch pie plate. In a blender or food processor, combine the sugar, eggs, cream, milk and vanilla extract and blend until smooth. Sift the flour over the mixture and pulse just to mix. Set the batter aside for 10 minutes.

While the batter is resting, peel the pears and cut them in half lengthwise. Using a spoon (a grapefruit spoon works best), remove the vein for the stem and the seed pit. Cut each half into thin crosswise slices between ¼ and ½ inch thick, keeping the pear in its original form. (This will take some time at first, but you'll soon get the hang of it.) As you finish each pear, flatten it slightly, lightly pushing down against the cutting board and toward the base of the pear. Lift it, using the flat of the knife as a spatula, and carefully place it in the pie plate with the stem end pointing toward the center. Depending on the size of the pears, only 5 of the 6 halves may fit. Pour the batter over the top and sprinkle with the pistachios. Sprinkle the top with sugar.

Place the pie plate on a baking sheet to catch any spills and bake until the clafouti is puffed and brown, about 45 minutes. Let cool slightly before serving.

CLAFOUTIS

The first time I saw a picture of a clafouti, I knew I had to make one. The dish was gorgeous, something like a cross between a pancake and a custard — puffed and brown and homey and dotted with melting cherries. I ran out and bought the cherries, whipped up the batter and stuck the clafouti in the oven. What came out was much more a pancake than a custard, and a pretty tough little pancake at that. *Oh, well,* I thought, and filed it away under "lessons learned."

Several years later, eating dinner with my daughter in a little French restaurant in New York, clafouti showed up on the menu. Still curious, I tried it again: it was nothing like the one I'd made. This clafouti was tender, almost custardy. It was perfect with cherries, and I could easily see how it could be adapted to fit other summer fruit.

This time I was determined to get it right. I gathered a stack of French cookbooks and went to work. Julia Child's recipe from *Mastering the Art of French Cooking* turned out to be much closer to what I had in mind. Lulu Peyraud and Richard Olney's *Lulu's Provençal Table* turned out to be much the same as Child's, though quite a bit sweeter (½ cup sugar to Child's ⅓ cup).

On a whim I checked out Joël Robuchon and Patricia Wells's *Simply French,* another cookbook that has been a never-fail source of great recipes. Predictably, Robuchon's recipe is very much a reinterpretation — a custard in a pastry crust, with fragments of cookie dough on top. But Wells included her own recipe — one made with pears and star anise. Surprisingly, the batter — made with equal parts cream and milk — turned out to be much more like what I had in mind than the milk-based recipes.

Armed with the fruits of my research and apricots and plums from the market, I set to work inventing my own clafouti. After several trials, I ended up with Wells's combination of milk and cream, but with more flour to make it a bit more cakey and much less sugar.

The truly wonderful thing about this recipe is its adaptability. A clafouti is just about the perfect way to present soft fruits. Tweak the seasoning just a little, and you can make this recipe with cherries (substitute cherry liqueur or vanilla for the almond extract), peaches or nectarines (combine with raspberries instead of almonds) or plums (a little ground cloves, or maybe dust the top with cinnamon sugar).

You can't imagine anything easier. Essentially, this is a very eggy pancake batter that you simply pour over sliced fruit. Mix the batter in a food processor or blender (the blender does a better job of dispersing the flour), let it stand for 10 minutes or so, pour it over the fruit and then stick the whole thing in the oven.

Pear Frangipane Tart

Frangipane is made by grinding nuts and eggs together to make something between a cookie and a custard. Use this basic recipe, varying the type of nuts and the flavorings, as a filling for all kinds of fruit tarts.

6 SERVINGS

> **Pastry for a 9-inch tart (see page 370)**
½ **pound blanched almonds**
⅔ **cup sugar, plus 1 teaspoon for sprinkling**
3 **large eggs**
1 **tablespoon oloroso or other sweet sherry**
2 **teaspoons vanilla extract**
¼ **teaspoon salt**
2 **tablespoons unsalted butter**
2 **teaspoons grated orange zest**
1 **tablespoon apple cider vinegar**
3 **Bartlett pears (½ pound each)**
1 **tablespoon unsalted butter, melted**

Prepare the pastry and fit it into a 9-inch tart pan with a removable bottom. Refrigerate.

Heat the oven to 375 degrees and place a baking sheet on a low rack. In a food processor, grind the almonds, ⅔ cup sugar, eggs, sherry, vanilla, salt, 2 tablespoons butter and the orange zest to make a smooth, slightly flowing paste.

Fill a large bowl about one-third full with water and add the apple cider vinegar. Peel the pears and cut them in half lengthwise. Using a spoon (a grapefruit spoon works best), remove the vein for the stem and the seed pit. As you finish each pear half, slip it into the bowl. Make sure there is enough water to cover all the pears.

Spread the almond mixture on the bottom of the pastry, using the back of a spoon to spread it evenly. Pat each pear half dry and carefully cut it into thin crosswise slices, keeping the pear in its original form. (This will take some time at first, but you'll get the hang of it quickly.) As you finish each pear, flatten it slightly, lightly pushing down against the cutting board and toward the base of the pear. Lift it, using the flat of the knife as a spatula, and carefully place it in the tart pan, with the narrow stem end toward the center. Place each subsequent pear slice next to the previous one in a spoke pattern until the tart is filled. Brush the pears with the melted butter and sprinkle with the remaining 1 teaspoon sugar.

Place the tart pan on the heated baking sheet and bake until the almond mixture is puffed and golden and the pears are tender, 30 to 35 minutes. Serve at room temperature.

Asian Pear Crisp with Walnut Topping

When cooked, Asian pears have a firmer texture than Western pears and a pronounced honeyed flavor, which is irresistible when paired with the smoky taste of bourbon.

6 SERVINGS

⅓ cup raisins

¼ cup bourbon, plus about 2 teaspoons for the cream

2 tablespoons walnut pieces

¼ cup all-purpose flour

2 tablespoons sugar, plus 1 teaspoon for the cream

¼ teaspoon salt

4 tablespoons (½ stick) cold unsalted butter, cubed

3 Asian pears (1½–1¾ pounds total)

2 tablespoons honey

1 tablespoon fresh lemon juice

⅓ cup heavy cream

Heat the oven to 350 degrees. Generously butter six ½-cup ramekins or baking cups. Place the raisins in a small bowl and add the bourbon. Let sit until the raisins soften slightly, about 30 minutes. Place the walnut pieces on a baking sheet and toast in the oven, stirring occasionally, until they are fragrant and have darkened slightly, about 15 minutes.

For the topping, combine the toasted walnuts, flour, sugar and salt in a food processor. Pulse 2 or 3 times to chop the walnuts coarsely; you don't want them finely ground. Scatter the butter over the dry ingredients. Pulse until the butter is in small clumps about the size of peas. Do not overprocess. Set aside.

Peel the pears and cut them in half lengthwise; then use a melon baller to scoop out the cores and stems. Cut the pears into neat ½-inch dice. You should have 3½ to 4 cups.

Place the diced pears in a large bowl and add the honey and lemon juice. Drain the raisins and add them to the bowl. Fold together gently to avoid breaking up the pears.

Spoon the pears into the prepared ramekins, filling them all the way to the top. (The filling will settle some during baking.) Scatter 1 heaping tablespoon of the topping over the pears. Do not pat firm — left a little loose, it will make a crisper crust.

Bake until the top is evenly browned, about 1 hour. Let cool slightly.

Meanwhile, whip the cream until it forms soft peaks. Beat in the remaining 1 teaspoon sugar and just enough bourbon to give the cream a faint flavor.

To serve, top each crisp with a small dollop of whipped cream and pass the rest of the cream at the table.

Persimmons and Figs

· · · · · · · · · · · · · · · · · · · ·

Most of fall's fruit harvest is straightforward enough and familiar to even the most oblivious of shoppers. But there are some oddballs hanging out in the produce department at this time of year, too — fruits with confused identities and even one with extremely unconventional sexual habits. Think of figs and persimmons as autumn's bohemians.

Persimmons are the confusing ones, because they come in two contradictory forms that to the untrained eye can look remarkably similar. Both are a deep harvest orange and are shaped a little like tomatoes. Both have a sweet, honeyed flavor. But as similar as they may seem, they must be handled in very different ways. One is so tannic that it must be ripened until it is almost custardy before it can be approached safely, at which point it is among the sweetest fruits on the planet. The other is at its best when it is crisp, but truth to tell, it can seem a little bland next to its kin.

The first type of persimmon is called the Hachiya, and when it is immature, it contains a tremendous concentration of dissolved tannins that are puckeringly astringent. You'll know this persimmon because it is shaped like a large acorn. Although it must be softened, you will usually buy it firm. (When it is ripe, it is so delicate that it will split like an overfilled water balloon.) Softening it takes no special trick, though. Just leave it at room temperature for a couple of days. The fruit is ready when it feels squishy (and if you're lucky, it may have developed some black streaking on the skin — this seems to go along with a really high sugar content). At this point, cut it open and spoon out the center. The pulp of a ripened Hachiya is about as sweet as any fruit you'll find. The suggested minimum sugar content for harvest is 21 to 23 percent. Find a peach that sweet, and you'll remember it all summer long.

Fuyu persimmons are smaller than Hachiyas (about the size of a pool ball rather than a baseball) and squatter. They look a little like miniature ottomans made from some exotic pumpkin-colored leather. Fuyus can be eaten right away. They have a pleasant, crisp flesh, and because they are less sweet than Hachiyas, they can be used in both savory and dessert dishes.

Hachiya and Fuyu are far from the only varieties of persimmons in the world, but because they are the most popular in the United States, these names serve as stand-ins for other lesser-known varieties that share the same traits as far as astringency and shape. (One group of botanists in Korea found more than 140 persimmon varieties.) Furthermore, the astringent acorn-shaped Hachiya group can be broken down into two subgroups — varieties that will sweeten on their own and those that must be pollinated first. There is a native American persimmon that is unrelated.

The two families of Asian persimmons came to the United States at different times and from different places. The Hachiyas were brought over by Japanese immigrants at the turn of the twentieth century. The Fuyus got their big push in the 1970s and 1980s, when a new wave of immigrants arrived from Southeast Asia, where they are the preferred variety.

Persimmons may be perplexing, but figs are downright kinky. Most of the fruits we eat are formed from the swollen fertilized bases of flowers. Reproduction follows a fairly standard plan: a brightly colored flower opens; pollinating insects (or birds, or even breezes) are drawn to the colors and alight on the pistils, spreading pollen; the fertilized fruit forms at the base of the flower; the flower drops away, leaving the fruit behind. Sex is not so conventional for the fig. The fruit that we eat is actually the flower — or, to be more accurate, cluster of flowers — turned inside out. To fertilize it, a special breed of bug — a tiny two-millimeter-long wasp — has to crawl up inside the fig, entering through a tiny hole at the base. It deposits its pollen on the thousands of tiny flowers that line the fig. (Those delicate little pops you feel when you eat a fig are actually hundreds of tiny seeds.) You don't find wasps or eggs in fresh figs at the market because the stems of the tiny flowers are long enough to prevent the females from laying their eggs, and any wasps that loiter too long are dissolved by a protein-digesting enzyme called ficin that is produced by figs. So how's that for kinky?

As you might expect, this complicated path to reproduction is littered with hazards. Chief among them in the United States is the fact that figs are a nonnative plant, and so that special wasp isn't native either. When immigrants from the Mediterranean brought their fig trees to California in the 1800s, they were puzzled as to why the trees never bore fruit. It wasn't until the latter part of the century that the mystery was solved. In the Old World, growers traditionally hung dried-out figs called "caprifigs" in their orchards every spring. Botanists put this down as a peasant superstition — until it was discovered that those caprifigs housed the wasps. It wasn't until a Fresno, California, farmer named George Roeding established the first colony of wasps in 1889 that commercial fig cultivation could begin on a grand scale in the United States. Today, when you drive up the Central Valley during the spring, you'll see fig trees hung with paper bags. These contain colonies of wasps, and this so-called caprification is the modern equivalent of the old tradition of hanging fig boughs.

Not all fig varieties require caprification, only the Smyrna group. One botanist determined that 471 varieties self-pollinate, producing fruit without the aid of the wasp. Not surprisingly, this group includes some of today's most widely grown varieties, such as Black Mission, Brown Turkey and Kadota.

The United States — or, more specifically, California, since that is where almost all of the crop is grown — is the second-largest grower of figs in the world, after Turkey. The vast majority of the harvest winds up being dried, most of it turned into paste that is used in cookies and other sweets. The very best figs are dried whole. Called "naturals," these account for a small percentage of the total, as the same wasp that pollinates the figs can also introduce fungi that cause spoilage during the drying process.

Fresh figs account for only about 9 percent of the California harvest, but they are a lucrative crop, bringing in as much as six or seven times the price of dried figs. Fig orchards bear two crops every year. First, in late spring or early summer, comes a small harvest of what are called "breba" figs. The second, or main, harvest comes in late summer and early fall. Because figs are so fragile — they actually make a Hachiya persimmon look hardy — harvesting them is a painstaking process. Each tree must be harvested carefully by workers wearing gloves and long sleeves, as the ficin that dissolves wasps is also extremely irritating to human skin. Each fruit has to be clipped from the tree, rather than being plucked from the branch as most fruits are. And to keep them from being squashed by their own weight, they are collected in little one-gallon containers rather than the usual big buckets. Finally, rather than being shipped to a main packing shed to be sorted into boxes, figs are packed right in the field, on wooden tables set up under tents. They go straight from the field to a refrigerated warehouse and then to the store. Each tree needs to be picked at least every other day during the harvest, which can last for several months.

Persimmons

· · · · · · · · · · ·

WHERE THEY'RE GROWN: Almost all commercial persimmons in the United States are grown in California, chiefly in the Central Valley.

HOW TO CHOOSE: Choose persimmons that are deeply colored. If picked too early, the orange may appear a little cloudy or pale. A streak of black on the skin of a Hachiya persimmon seems to indicate an especially sweet piece of fruit. Choose Fuyu persimmons that are uniformly firm and have no blemishes.

HOW TO STORE: Persimmons must be stored at room temperature. If refrigerated, they will quickly develop chilling damage. Store unripe Hachiya persimmons in a closed paper bag with an apple or pear — the ethylene gas produced will speed ripening. Freezing overnight will soften Hachiya persimmons (and moderate the tannins), but somehow the flavor is never quite as good.

HOW TO PREPARE: With ripened Hachiyas, just cut them open and spoon out the pulp. With Fuyu persimmons, just cut into wedges and remove the few large central seeds.

ONE SIMPLE DISH: For Hachiyas, cut a shallow X through the tip of the fruit and peel back the skin, then crown it with a little bourbon-flavored whipped cream and some toasted chopped walnuts or pecans. Fuyus are good cut into segments and macerated with a little sugar and some orange-flavored liqueur. Top with toasted sliced almonds.

Figs

• • • •

WHERE THEY'RE GROWN: Almost all of the commercial fig harvest — fresh and dried — comes from California, mainly the Central Valley.

HOW TO CHOOSE: Fresh figs are incredibly fragile when ripe, and since they don't improve after picking, that's the only way you should consider buying them. Look for the softest figs you can find. Don't worry if there are some slits and tears in the skin, as long as the fig smells fresh and fragrant. Figs that are overripe and have begun to spoil will smell frankly of fermentation.

HOW TO STORE: Figs are fine in the refrigerator. In fact, they are so delicate that they really must be chilled — they can begin to spoil after only a couple of hours of warmth.

HOW TO PREPARE: There are those who peel figs, and this may not be such a bad idea if they have thick skins or scab spots. In general, however, peeling is not necessary. About the only preparation needed for a fresh fig is removing the tiny bit of stem.

ONE SIMPLE DISH: There is no better fall dessert than quartered fresh figs macerated with sugar and a little rum and spooned over good vanilla ice cream.

Fuyu Persimmon Salad with Cumin-Lime Vinaigrette

Because Fuyus are only slightly sweet, they make a refreshing salad when paired with an assertive vinaigrette such as this one. But dress them with just a little lime juice and a teaspoon or so of sugar, and you have a great fall dessert to finish a big meal.

8 SERVINGS

- 2 pounds Fuyu persimmons
  Juice of 1 lime, plus more to taste
- 2 tablespoons walnut oil
- ½ serrano chile, seeded and minced
- ½ teaspoon ground cumin
  Salt
- ¼ cup pomegranate seeds (about ½ pomegranate)
- 3 tablespoons chopped walnuts, toasted
- 2 tablespoons chopped fresh cilantro

Cut off the tough green calyxes and slice each persimmon into 10 to 12 wedges.

In a small lidded jar, combine the lime juice, walnut oil, about half of the chile, the cumin and a dash of salt. Tightly cover and shake hard to mix well. Taste the dressing on a small piece of persimmon. There should be just enough chile to add a suggestion of heat. If you'd like it hotter, add more and shake again.

Combine the persimmons and the dressing in a bowl and toss to coat well. Turn the persimmons out into a decorative bowl and sprinkle the pomegranate seeds, walnuts and cilantro on top. Taste and add more salt or lime juice, if necessary. Serve.

Fig-Honey Gelato

Figs add a lovely luxurious quality to any dish in which they are used. When you taste this gelato, you won't believe that it's made with almost all milk. Figs contain an enzyme called ficin, which can turn milk sour. Heating the fruit and sugar to 155 degrees kills the enzyme.

MAKES 1 QUART

- 1 **pound fresh figs, quartered**
- ½ **cup sugar**
- 1½ **cups milk**
- 2 **tablespoons honey**
- 1 **tablespoon orange liqueur**
- ⅓ **cup mascarpone**
- **Pinch salt**

Heat the figs and sugar in a saucepan over medium heat, stirring roughly so the figs break apart. Cook, stirring, until the figs have mostly melted and just begin to bubble, about 5 minutes.

Remove from the heat and stir in the milk, honey, orange liqueur, mascarpone and salt. Chill well, then freeze in an ice-cream machine according to the manufacturer's instructions.

Market Corrections

THE RETURN OF THE SMALL FARMER

Although the Weiser family of Tehachapi, California, may have its quirky members, none of them could really be considered revolutionaries. And when they started selling their produce at farmers' markets back in the early 1980s, they weren't trying to change the world; they just wanted to get what they thought was a fair return on their labor.

Even so, what began as the agrarian dream of a former high school guidance counselor is now a business that sells its fruits and vegetables all over the United States and grosses more than $1 million a year. But if that big-money talk makes you think of gentlemen farmers wheeling and dealing over the long-distance lines, forget it. You can still find one of the sons, Alex Weiser — small and wiry and dressed in jeans and a T-shirt — at several of the more than a dozen farmers' markets around Southern California every week. Indeed, the Weiser family represents one of the most profound transformations in American agriculture in the past twenty-five years — the return of the small family farm as a viable way of doing business.

Just when it seemed that modern farms had to get bigger or die, along came farmers' markets to save the day. By going to the market, growers were able to realize the full cash value of their produce, rather than the measly 20 percent or less that was the norm in mainstream agriculture. The markets also allowed farmers to experiment with new crops that might not have fit into the supermarket produce section and even paid them a premium for doing that. In so doing, farmers' markets shook the foundations of the produce industry by unleashing consumer demand for fruits and vegetables that were more varied and more flavorful than those that had been offered before and demonstrated that shoppers were willing to pay for them. But the revolution hardly stopped there.

The changes that these markets introduced also had an effect on the markets themselves. What once were simple gatherings that accomplished nothing more than allowing farmers to sell directly to their customers evolved into something much more vibrant and complex. Ultimately, the changes altered the way the growers who sold at the markets farmed, even down to the choices of the crops they grew.

In the Weisers' case, selling directly to consumers is no longer anywhere near their biggest source of income, but their presence at farmers' markets remains the most important factor in their success. The market stands give the family the credibility to charge higher than commodity prices even for fruits and vegetables they don't sell at the markets. By growing good food and then getting good prices for it, the Weisers have been able to support four families and more than a dozen employees on only a little more than a hundred acres. Still, they are quick to point out that working farmers' markets is only marginally more profitable than selling to wholesale accounts. They say that their farmers'-market expenses range from 70 to 75 percent, and it takes a heckuva lot of work — certainly more than loading up the truck and driving to the packing shed.

Using farmers' markets as a springboard to greater things was not the Weisers' intent when they started. It all developed pretty naturally. The family got into agriculture in 1976, when the father, Sid, retired from his job as a gang counselor at Garfield High School in East Los Angeles. Sid had always wanted to farm, so when he learned of a 160-acre apple orchard for sale, he jumped on it. He soon wished he hadn't. "After about five years, we nearly went bankrupt," he says. Alex, the son who runs the farming end of the operation these days, remembers his first experiences trying to sell their apples through the normal commodity route: "We'd take them to the packing shed, and they'd beat you down for a buck. We were really passionate about what we were growing, and they thought that was kind of funny." Every grower who has ever sold at a farmers' market has some version of the same story. The Weisers started going to the farmers' markets as an experiment to see if they could earn some quick money to tide them over between commercial sales. The strategy worked. "Going the regular way, we might not make enough money to cover the costs of packing and production," Alex remembers. "But I was bringing in cash from the markets."

Eventually, the Weisers ripped out the apple trees they'd started with (even at farmers' markets, it's tough to compete with the flood of fruit from the Northwest). Today the Weisers grow dozens of fruits and vegetables. Melons and potatoes are their specialties, and they produce dozens of varieties of each. "The markets allow farmers to find niches," Alex says.

His ideas for new things to plant come from a variety of sources. He is constantly scouring seed catalogs and Web sites ("farmer porn," one family member calls them), and the markets work not only as a place to sell crops but also as a source of new ideas. Sometimes Alex will see things at a competitor's stand that he wants to try. For instance, during the spring lilacs are a big moneymaker at farmers' markets. The Weisers had always had lilacs on the farm, but it wasn't until they saw the lines forming at another vendor's stand that they thought about selling them.

More frequently, a customer will ask about a fruit or vegetable she has bought someplace else or remembers from her childhood. And quite often chefs will make special requests. At the begging of one regular restaurant customer, Weiser planted something called *crosnes,* a knotted white tuber that is all the rage in France. Other farmers have even gone so far as to venture out into the hillsides to dig wild stinging nettles, arugula and fennel when their customers have asked for them.

The restaurant connection is one of the most important links in the Weiser family's market chain, and it was established slowly and almost accidentally. The number of chefs who shop at farmers' markets is relatively few, but they can make a significant difference to small farmers. First, they buy in bulk. It is much easier to sell eighty bunches of beets to one customer than one bunch to eighty people. Because chefs buy in such large quantities (and because, unlike home shoppers, they really need those beets if they have put them on the menu), the Weisers began to let them call in their orders in advance to reserve their purchases.

The next step was subtle but critical: some chefs who couldn't make it to the market every week started asking the Weisers to deliver to their restaurants. The family didn't have the trucks or the manpower to do that, so they arranged with the restaurants' normal produce purveyors to make the deliveries. And then those purveyors started asking about picking up a little extra to sell to their other clients.

Thus, without really trying to, the Weisers got back in the wholesale business. But this time, because of the reputation they had earned by working the markets, they were able to get well above standard wholesale prices for their crops. Indeed, Alex points out, sometimes their wholesale customers are willing to pay even more than the folks at the markets. Pretty soon the Weisers were selling to a dozen different wholesale companies — a couple of them with national distribution — and their produce began showing up at high-end supermarkets around the country. This is not unusual. According to one survey, more than 70 percent of growers who sell at farmers' markets also sell though traditional channels.

Now, although the markets are at the core of the Weisers' business, sales outside the markets make up the bulk of the farm's income. Dan Weiser, Alex's brother and the business brain in the family, estimates that sales to restaurants and wholesalers now account for 60 to 65 percent of the farm's business. And, he says, he'd like to take that number to 90 percent, an idea that makes Alex noticeably uncomfortable.

Although Alex acknowledges the need to expand the business, he insists that farmers' markets and sales to the public should always be at the center of the operation. "The market gives us the confidence we need to grow stuff," he says. "It confirms our ideas about what people are looking for. Also, knowing that I can always sell stuff here gives me leverage I wouldn't have otherwise when dealing with produce companies. If they can't give me a decent price, I know I can move it myself rather than just taking what they want to give me. I know I'm not going to lose my shirt on something; I've always got my costs covered. Farmers' markets give us a little control over our fates."

Furthermore, the markets act as magnets for other business. Specialty produce companies regularly troll farmers' markets looking for new items and suppliers. "It's kind of a showroom for new products," Alex says. Executives from big specialty produce wholesalers often come by the stand, whip out digital cameras and snap pictures of the vegetables on display. Dan, who joined the family business a couple of years ago after working in entertainment marketing, says, "The word we used for it at Disney was 'synergy.' We couldn't do restaurants or wholesale without the farmers' markets. And if we were doing just farmers' markets, we wouldn't be able to make enough money to keep going."

Of course, not every farmers' market grower is fronting a million-dollar business. Many are just scraping along. But sometimes even that is a marked improvement for the farmers in question, and some of them have had a significant effect on the produce industry. There is no better example of that than the more than seven hundred Asian American farmers, predominantly Laotian, who live in Fresno County, California. More than half of the certified farmers' market growers in the area — the richest fruit and vegetable farmland in the country — are of Asian descent. Although their economic impact is hard to quantify, the statistical category "Asian Vegetables" accounts for more than $10 million in sales in Fresno County alone, a figure that is growing every year.

Many of the Laotian farmers came to the United States after the Vietnam War. Some had fought in the United States "secret war" in Southeast Asia and suffered years of hardship after the Americans left the region in 1973. Christian groups throughout the United States sponsored the immigration of many Laotians at that time, and gradually a large number of these people relocated to Fresno. Here they live, frequently in grinding poverty, as small farmers. Against all odds, they are not only succeeding in feeding their families, but they are also changing the way we eat. Almost every farmers' market in California has at least one Laotian farm stand. Sometimes the Laotians sell standard American fruits and vegetables with which they were familiar in their homeland. Cherry tomatoes, eggplants and peppers are mainstays. Often, however, the stands are stocked with a wealth of produce that, to American eyes, seems quite exotic: green and purple yard-long beans; bright orange melons that look like spiny cucumbers and have huge pomegranate-red seeds; squash that can be eaten like zucchini when they're young or used as a bath sponge when they mature; stalky Chinese broccoli; giant white daikon radishes; eggplants of every color and shape; water spinach; exotic mints, basils and other herbs; and the tender green shoots of pea plants.

So numerous are these growers that the local morning farm reports are now broadcast in the Hmong language by a Hmong county extension agent named Michael Yang. (Many of the Laotians are Hmong, a group that lives in the mountains of northern Laos.) A short, slightly stocky man with a quiet demeanor, Yang seems much like any other county extension agent except that when he steps into the sun, he pulls on a conical straw hat instead of a cap with a seed company logo. Yang's family immigrated to the United States in 1980, when he was eight years old. His father had worked for the Americans in Laos. After he was killed, it fell to Michael, the oldest son, to guide his mother and brothers and sisters through the jungle to Thailand and the refugee camps there. At one point he was bitten by a two-foot-long centipede and was certain he was going to die. His mother carried him on her back for three days until he could walk again.

Much of Yang's work is done at the Hmong American Community educational farm. The twenty-acre plot is divided in half, with separate fields for the lowland Lao (another Laotian people) and mountain Hmong farmers. Partly the separation is necessary because the two groups prefer different vegetables, but there is well-documented friction between the groups as well. The Hmong side is planted with cool-weather leafy vegetables and herbs that grow through the winter: bok choy, Chinese broccoli, water spinach and napa cabbage, as well as various kinds of mint and basil. The Lao side is covered with trellises garlanded with bitter melon, yard-long beans and loofah. Long, lavender Chinese eggplants and small, round, green Thai ones grow in rows.

New vegetables are added every season, some from families' private seed collections. Some items are so obscure that Yang says they have to be sent to the University of California at Davis for genotyping to be identified. There is also a financial imperative to their experimentation. Even as the demand for Asian ingredients grows, small farmers find it difficult to compete against larger operations in the United States and Mexico, so the Laotian growers are constantly looking for an edge with niche products.

Most of the produce the Laotian farmers grow goes through normal distribution channels, but much of it also is sold at farmers' markets up and down the state. Fresno farmers travel as far as San Diego and San Francisco — up to six hours each way — to sell their crops. At wholesale a 30-pound case of eggplants might sell for $6. At the farmers' market growers can get $1 a pound. This pays enough that one of the farmers has put three sons through college on less than 25 acres of mostly jujubes (Chinese dates) and daikon.

A lot of this produce is so exotic that you probably won't find it at your local supermarket or chain restaurant today, but who knows what tomorrow might bring? Baby bok choy and daikon, which are everywhere today, were considered exotic only five years ago. Can Chinese broccoli, yard-long beans and fuzzy melon be far behind? The mainstream demand for these ingredients started with adventurous chefs looking for new things to play with, but many of them have been picked up by chain restaurants such as Applebee's and Bennigan's as they flirt with Americans' taste for Asian food. This quiet introduction of new ingredients is guerrilla marketing at its best. The specialty produce distributor Frieda's, which introduced America to the kiwifruit, now has a list of more than thirty Asian vegetables, including arrowroot and yu choy sum, a kind of flowering, mustardy green.

Mainstream acceptance is the Laotians' dream for tomorrow. For now, bouncing around Fresno in his dusty 1983 Toyota, the air conditioner straining against the 106-degree heat, Yang is focused on the present and helping other Laotians attain the success he enjoys. After more than twenty years in this country, he says he has finally stopped having nightmares about the jungle and his family's exodus. He turns down a dirt road between brand-new residential developments and stops beside a verdant plot, not more than an acre and a half. At the far end are fat stalks of sugarcane, elephant ear leaves of taro and what looks like tall grass. That grass is rice, a specially prized variety traditionally grown in the Laotian highlands. In another month it will be harvested and sold for a very good price at one of the farmers' markets, or to a restaurant. Who knows? Maybe someday you'll even be able to buy it at your neighborhood grocery. But now it sits baking in the Fresno sun, giving off a heavenly aroma that smells to an outsider like the very best basmati. For a Hmong farmer, the fragrance is like the past and the future combined. Yang stands at the edge of the field, closes his eyes and breathes in, deeply, over and over again.

Winter

Cabbages and Brussels Sprouts

. . . .

At one time not so very long ago, the Brassica family lived in the ghetto of the vegetable kingdom, never allowed out in polite company. Presidents could joke with impunity about hating them. But a funny thing has happened: mustard greens have gone trendy. Broccoli is suddenly fashionable as a pasta sauce. Cauliflower is used at fancy restaurants as a complement to caviar (a role in which it excels). Kale is grown as a garden flower. Even so, while many of their brothers and sisters are attaining stardom, two members of the far-flung clan seem to be makeover-proof. Cabbages and brussels sprouts are the black-sheep brassicas.

What an unfair rap for such sweet kids, because when they are carefully cooked, no vegetables are more delicious. Cabbage can be silky and sweet, with just a slight bitter edge from the brassicas' characteristic mustard oil. Brussels sprouts can be sweet as well, but in addition to the mustard overtones, they have an intriguing earthy flavor. On top of that, they are absolutely gorgeous, with buttery yellow inner leaves contrasting with forest green outer ones.

Unfortunately, they frequently don't get the care they deserve. And when that happens, these brassicas bite back. Indeed, the very word "cabbage" conjures up the smell of Irish tenement hallways. If anything, brussels sprouts' reputation may be even worse. Not only do they develop that distinctive smell when overcooked, but their colors fade to a uniform shade of olive drab.

Like all brassicas, cabbages and brussels sprouts are high in chemical compounds that produce hydrogen sulfide when exposed to heat for a sufficient amount of time. (As a general rule, any chemical compound with any variation of the word "sulfur" in it smells bad.) Brussels sprouts produce twice as much of these sulfurous compounds as broccoli, and green cabbage is not far behind. This chemical is produced by a process similar to that which occurs in onions: chemicals that are separate in the raw vegetable combine and form new compounds when the cell walls are damaged. In this case, the compounds form when the leaves are chopped and especially when the cell walls are broken during heating. And unlike the sulfurous compounds in onions, those in brassicas don't dissipate as cooking progresses, but instead increase, as more cell walls soften and rupture.

Overcooking is the bane of all brassicas, but it's a particular problem with brussels sprouts. In the first place, most people don't cook them often enough to gain experience with them. In the second place, because they're such dense little cabbages, even good cooks sometimes feel the need to overcook them to tenderize them.

To treat brussels sprouts with respect, begin by shopping for the smallest sprouts you can find. These will cook the fastest and have the sweetest flavor. When you're getting them ready for cooking, be sure to remove any dark or damaged outer leaves and trim away the dark, dried-out base of each sprout. Cut an X through the base ¼ to ½ inch deep, depending on the size of the sprout. This will allow the heat to penetrate to the heart (where the offending chemicals are concentrated) but still hold the sprout together so you don't wind up with a lot of loose leaves. When you cook the sprouts, never cook them for more than 7 minutes during the initial heating. After that, you can actually smell the change from sweet cabbage to sulfur begin. If you would rather blanch the sprouts in boiling water, make sure there is plenty of it, to dilute any acids given off during cooking.

With cabbages, the solution is even easier. When preparing dishes using raw chopped cabbage, simply soak the vegetable in cold water after cutting it up. This not only revivifies it and makes it crisper, but it also leaches out many of the chemicals that turn sulfurous. These compounds are also reduced when cabbage is fermented into a pickle, as in sauerkraut or kimchi. Boil cabbage briefly to reduce the amount of sulfur, and cook it in plenty of water to dilute any that is created.

The thick, dense leaves have a slightly waxy texture, and these characteristics account for much of the vegetable's enduring appeal. Because the leaves are so thick, they make good containers and are easy to stuff, whether they have been blanched and softened (in the well-known European fashion) or left raw and crisp (as in the less familiar Asian way of using cabbage leaves as cups or wrappers for chopped mixtures). Because of the leaves' density, when cooked they are at once sturdy and silky. Add ribbons of cabbage to a soup, for example, and they will not only sweeten the broth but also turn the dish into a sort of vegetable pasta. Combine that resiliency with the slightly waxy texture of the leaves, and you understand why cabbage is the premier ingredient for salads such as coleslaw. Whereas more delicate greens must be dressed at the last minute to avoid wilting, cabbage salads actually improve as they sit.

There are many different kinds of cabbages, but they tend to fall into three general categories: round, red and long-headed. Round cabbages are the most common and include the familiar smooth-leaf white varieties as well as the crinkly-leaf green Savoy types. These normally grow to a size somewhere between a softball and a volleyball, but some varieties can swell to more than 100 pounds. Among the hardy gardening subgenre that competes to grow the most outsize specimens, cabbages have a place of honor, along with pumpkins, watermelons and zucchini. The world record, held by a Welshman, is 124 pounds.

Red cabbages look almost exactly like round cabbages except they are tinged with anthocyanin, a pigment that lends a surface color somewhere between red and purple. This color fades during cooking to a bruised blue. Cooking the cabbage with acidic ingredients such as apples or vinegar can lessen this change.

The most widely known long-headed Asian cabbage is the napa (or nappa; the name derives not from the wine country, but from the Japanese word for cabbage). These cabbages form pale yellow-green heads that are elongated rather than round, much the same as the difference between romaine and iceberg lettuces. Asian cabbages (*Brassica rapa*) actually come from a different species than European cabbages (*Brassica oleracea*). They are more closely related to bok choy, broccoli rabe and, most oddly, turnips. Asian cabbages also are more delicate in texture and flavor than European cabbages.

WHERE THEY'RE GROWN: Cabbages and brussels sprouts require long, cool growing conditions. For that reason, they are usually at their best beginning in late fall and running through early spring. Essentially all brussels sprouts grown in the United States come from California, primarily the coastal region between Salinas and Santa Cruz. California is also the leading grower of cabbages, with about 20 percent of the crop. Cabbage is also an important crop in New York, as well as in Texas and Florida, where it is a winter harvest.

HOW TO CHOOSE: Choose brussels sprouts and cabbages in the same way — the difference is scale. The first thing to look for is tightly formed heads. Squeeze the head — there should be little give. Avoid any heads with discolored or damaged leaves or stems that appear dried out. Produce managers frequently "tidy up" cabbages and brussels sprouts by trimming the bases and discarding any faded leaves. Avoid any that look as if they've been overworked.

HOW TO STORE: Store cabbages and brussels sprouts in plastic bags in the crisper drawer of the refrigerator. They're tough enough to last a long time, but they're at their best within the first couple of days after purchase.

HOW TO PREPARE: For brussels sprouts, trim the base and discard any fading leaves. Cut an X in the base and either steam or blanch the sprouts just until tender, 5 to 7 minutes. For cabbages, discard any fading leaves, cut into vertical quarters and trim the solid core from the center. (If you want the cabbage wedges to remain whole, skip this step.) Steam or blanch just until tender, 7 to 10 minutes. If you're going to use the cabbage raw, as in a salad, shred it and soak it in ice water for 5 minutes to crisp it and remove some of the mustardy flavor.

ONE SIMPLE DISH: *For brussels sprouts:* Steam as above, then cut each sprout lengthwise into quarters. While the sprouts are steaming, chop 2 or 3 slices of bacon and render in a skillet until crisp. Add ½ cup red wine vinegar, raise the heat to high and cook until the vinegar loses its harsh smell. Reduce the heat to low and add the brussels sprouts, cooking and tossing until they are heated through, about 10 minutes. Sprinkle with ¼ cup toasted pine nuts.

For cabbages: This works best with Asian cabbages. Shred the cabbage thinly and soak in ice water. Pat dry. In a large bowl, whisk together 1 teaspoon minced garlic, 1 tablespoon vegetable oil and 1 tablespoon rice vinegar. Add the cabbage, season with salt to taste and toss to combine. Toss with torn cilantro leaves.

Lentil Soup with Sausage and Cabbage

Adding the shredded cabbage at the end lets it melt into this earthy winter soup, providing a subtle sweetness. Don't neglect the final step of adding a hint of acidity — it's surprising how much this can pull the flavors into focus.

8 SERVINGS

- 2 **tablespoons olive oil**
- ½ **pound Italian sausages**
- ½ **pound beef stew meat (1-inch cubes)**
- ¼ **cup red wine vinegar, plus more to taste**
- 2 **carrots, chopped**
- 2 **celery stalks, chopped**
- 1 **onion, chopped**
- 3 **garlic cloves, chopped**
- 1 **pound lentils**
- 10 **cups water**
- 1 **bay leaf**
- 1 **28-ounce can crushed tomatoes**
- **Salt**
- 1 **pound cabbage**
- **Freshly ground pepper**

Heat the oil in a large, heavy pot over medium heat. Slit the sausage casings and crumble the meat into the pot. Discard the casings and break the meat up with a spoon. Chop the stew meat as finely as possible by hand and add it to the pot. Cook, stirring occasionally, until the meats are well browned, about 15 minutes. Be careful not to let them scorch.

Add the vinegar and stir; it will evaporate almost immediately. Using a slotted spoon, transfer the meat to a strainer and shake it to remove as much grease as possible. Discard all but 1 tablespoon of the grease in the pot.

While the meat is draining, add the carrots, celery and onion to the pot and return it to medium heat. Cook, stirring frequently, until the vegetables soften and become fragrant, about 5 minutes. Add the garlic and cook until fragrant, 2 to 3 minutes more.

Add the lentils and stir to coat with the flavorings. Add the meat, water, bay leaf, tomatoes and 1 teaspoon salt and bring to a simmer. Cook until the lentils have softened, about 45 minutes.

Cut the cabbage into quarters. Trim out the core and shred the cabbage coarsely. About 15 minutes before serving, add the cabbage to the pot and stir to mix well. Let the cabbage melt into the soup, then season to taste with a good grinding of pepper and more salt and vinegar, if necessary. Remove the bay leaf and serve.

Duxelles-Stuffed Savoy Cabbage

This recipe comes from my friend the brilliant French chef Michel Richard. He frequently uses curry powder as a seasoning for mushrooms. In the quantity called for here, you don't so much taste it as notice the complexity it brings to the mushrooms. If you don't have any hazelnut oil, you can use olive oil instead.

MAKES 12 ROLLS

- 1–3 heads Savoy cabbage (enough for 12 large leaves)
- 1 tablespoon hazelnut oil or olive oil
- 2 pounds mixed mushrooms (button, shiitake and portabello), finely chopped
- 1 teaspoon curry powder
- Salt and freshly ground pepper
- 1 tablespoon olive oil, plus more for the baking dish
- 1 onion, minced
- 2 garlic cloves, chopped
- 2 cups chicken broth

Remove the outer leaves of the cabbage, choosing 12 of the largest. Blanch the leaves in plenty of rapidly boiling salted water just until they soften, about 2 minutes. Immediately remove them from the water and shock them in a bowl of ice water to stop the cooking. Press them dry between paper towels. Cut away the thick white spine from the center of each leaf and set the leaves aside.

Chop enough of the inner cabbage leaves to make 2 cups.

Heat the hazelnut oil in a large skillet over medium heat. When it is hot, add the mushrooms and curry powder and cook, stirring, until the mushrooms soften, about 5 minutes. Season with ¾ teaspoon salt and ¼ teaspoon pepper, or to taste, and set aside to cool briefly.

Heat the oven to 325 degrees and grease a 9-by-13-inch glass baking dish with olive oil. Spread one of the blanched cabbage leaves flat on a work surface with the outside of the leaf facing down. Season it lightly with salt and pepper. Spoon about 3 tablespoons of the cooked mushroom mixture into the middle of the leaf and wrap the leaf around it to make a little package. Place it seam side down in the baking dish. Repeat until you have 12 little packages in the baking dish. Wipe out the skillet.

Heat the olive oil in the skillet. Add the onion and cook until softened, about 5 minutes. Add the garlic, chopped cabbage and chicken broth and bring to a simmer. Cover and cook until fragrant, about 20 minutes.

Let the mixture cool slightly. Ladle the mixture into a food processor and puree it, then pour it over the cabbage rolls. Bake, uncovered, until the rolls become firm, about 1 hour, occasionally spooning the cooking juices over the top. Serve.

Cooking Greens

· · · · · · · · · · · · · · ·

Some vegetables are born to be stars; others are better suited to ensemble roles. There is no better example of the latter than winter's hardy cooking greens. While you can cook mustards, chards, collards, kale and even the leafy parts of turnips and beets individually, they are best prepared en masse, or, as your mama might have said, "in a mess." Taken one at a time, each of these greens has something to recommend it. But cook them together, and the result is extraordinary. The flavor of mixed greens is full and deep, rather than sharp and pointed. Cook one type of green, and you have a string quartet. Cook a mess, and you have an orchestra.

Moreover, a carefully prepared mix of greens is not only delicious; it has the power to heal. If you're feeling beaten down, peckish, fluish or even just a little mulish, nothing restores your equilibrium like a bowl of greens. "What is patriotism but the love of the good things we ate in our childhood?" Lin Yutang famously observed, and the same could be said for tonics. If you were raised in the South, when you're in need of sustenance, everyone else can keep their chicken soup.

Most greens have a deep, slightly sweet flavor with a wonderfully biting bitter backbone. In the South it's traditional to simmer greens for a couple of hours with a good-size chunk of fatty pork. This is the kind of perfectly realized rustic dish that if it had originated in, say, Liguria or Provence would now be offered in every upscale trattoria and bistro between Berkeley and Manhattan. But that's hardly the only thing greens are good for. They make a great bed for cooking Italian sausages. Or you can add greens to your favorite soup or stew a few minutes before serving for a vibrant bit of color and texture. They are even surprisingly at home in elegant surroundings. Fold them into a soufflé, for example, or bake them in a tart.

It used to be that greens could be found only in specialty markets in certain neighborhoods. Now they're in every upscale grocery. Can gourmet bacon grease mixed in with all those extra-virgin olive oils be far behind? And with the current trend toward precut, "preprepared" produce, you can even find cellophane bags of cooking greens, both individually packed and mixed, that have already been washed and chopped. These are not quite ready to cook — a machine can never do as good a job of sorting and trimming out stems as a human being — but they are still a convenience.

Although most southern greens may be best as a collective, each variety is slightly different, and good cooks are pretty picky about the perfect ratio of one type to another. There is nothing haphazard about a well-prepared mess.

Mustard greens are the most fragile. Their flavor is strong and peppery, but their texture is almost frilly, and their color is pale. In greens-loving neighborhoods, you'll sometimes see a distinction made between mustards and "Texas mustards," which are even more delicate. These can even be used in salads, albeit sparingly, as you would a strong-flavored herb.

Collards are the sturdiest greens, with a leathery texture that takes some cooking to break down. Once they are cooked thoroughly, however, their texture is downright silky, and their flavor is complex and minerally.

Kale also is tough. It's the green that's used for garnish in cheap salad bars because it can be left for days without wilting. It is usually the darkest green in the bunch, and both its flavor and cooked texture can best be described as meaty. Particularly good is the old-fashioned Tuscan *cavalo nero* or *lacinato*, made over for a modern mass audience and relabeled "dinosaur kale" because of its rough texture. Its color is so dark that it's nearly black, but when cooked it brightens to a brilliant forest green. Cook it low and slow, and the flavor will be downright sweet.

Some greens are the leaves of plants that are primarily grown for other parts. Beet and turnip greens taste true to their roots and are among the tenderest of all greens.

And then there are the Asian cooking greens. The most familiar are bok choy and the slightly mustardy green commonly called Chinese broccoli, Chinese kale or gai lan. Bok choy generally forms a very large head with dark green leaves and stark white ribs. There are a couple of smaller varieties, too. Cantonese bok choy looks the same, but in miniature; Shanghai is small and has jade green ribs; Taiwanese is smaller and leaner. Bok choy is generally very mild-tasting and is good simply steamed or blanched. Chinese broccoli is probably a distant relative of the Mediterranean varieties, but it consists mainly of thin stems and leaves. It is best blanched and sautéed. Take a hint from the flavor similarity and use it as a substitute for the much more expensive broccoli rabe in Italian dishes.

Finally, there are the chards, which are appreciated as much for their stems as for their leaves. Most commonly, you'll find red- and white-stemmed chards, but there is also a new variety with stems of varied colors called rainbow chard, which is sometimes packaged under the *nom de commerce* Bright Lights. The flavor of chard is minerally, with a beetlike sweetness that differs depending on the variety. Perhaps predictably, the redder the stems, the stronger this flavor will be. (The red hue comes from betalain, the same pigment that colors beets.) Chard leaves are nearly as leathery as collard greens. Still, they are far tenderer than the stems and should be cooked separately, at least at first.

Southern cooks most commonly combine all of the greens and boil together, but because of their varying textures, it's often a good idea to start them separately, particularly if you're going to sauté them afterward.

Blanch the greens in separate batches in a big pot of rapidly boiling salted water just until they wilt and become tender. Mustards will be done almost instantly. Kale will take 3 to 4 minutes. Chard leaves will take 4 to 5 minutes (the stems will take 5 to 7 minutes). Collards can take nearly 10 minutes. Once you've blanched them, combine them to finish the cooking.

Blanching is necessary only when you want all of the greens to have the same texture. Sometimes that's not desirable. In the ricotta tart on page 336, for example, the variety of textures is part of the attraction. Some of the greens are chewy; others are almost melting. In Southern Comfort Soup (page 335), blanching isn't necessary because the greens are minced so finely in the food processor that the differences in texture aren't really noticeable.

Perhaps paradoxically, before the color fades, it first becomes much more intense. Time it just right, and you can end the blanching just at that moment, preferably by shocking the greens in an ice-water bath to stop the cooking cold. This brightening happens because the cooking drives off the air that is trapped in the leaves along with the chlorophyll. With the air gone, the bright green color shines much more vividly.

WHERE THEY'RE GROWN: More than half of the collards in our markets come from Georgia, and more than half of the kale comes from California. California grows more mustard greens than Georgia, North Carolina, South Carolina or Texas.

HOW TO CHOOSE: Greens should be rigid and firm, not wilting. Avoid greens with black spots on the leaves, which are a sign of breakdown.

HOW TO STORE: Greens should be refrigerated immediately in a humid environment, such as the crisper drawer. Keep them in plastic bags, but if you're going to store them for very long, slip in a paper towel to absorb any condensation.

HOW TO PREPARE: Be sure to wash greens well, especially those with crinkly leaves, such as mustards — there are plenty of places for sand and soil to hide. The stems of most greens should be removed before cooking. They are thick and tough and will soften much more slowly than the leaves (although if you're going the southern route — cooking greens forever — this is much less important). Once you've removed the stems, the greens can be stacked and shredded quite easily.

ONE SIMPLE DISH: I serve braised greens as a side dish to roasts all through the winter. Render a little pork fat in a deep-sided sauté pan — bacon, pancetta, even prosciutto will do nicely. Clean the greens, removing the stems. Wash the leaves well and toss them by large handfuls into the pan with plenty of water still clinging to the leaves. As one batch cooks down, add another. Finish with a little sherry vinegar for backbone.

Southern Comfort Soup

There is something inordinately comforting about a big bowl of long-cooked greens.

6 SERVINGS

  2 tablespoons olive oil
  4 garlic cloves, minced
  ½ pound mixed leafy greens without stems (mustard, kale, collard, beet, turnip and chard)
  6 cups weak vegetable or chicken broth
    Salt
  1¼ cups water
  ¾ cup jasmine rice
  1½ teaspoons sherry vinegar, plus more to taste
    Freshly ground pepper
    Freshly grated Parmigiano-Reggiano

Heat the olive oil and garlic in a soup pot over medium-high heat until the garlic softens, about 3 minutes. Coarsely chop the greens and add them to the pot. They will come close to overfilling, but within about 5 minutes of cooking and stirring they will wilt down to almost nothing. Add the broth and 2 teaspoons salt and slowly bring to a simmer.

While the greens are cooking, heat the water and rice in a 1-quart sauce-pan over medium-high heat. When the water comes to a boil, reduce the heat to low and cover tightly. Cook until the bottom of the pan is dry and the rice is tender, about 15 minutes. Remove from the heat and let cool, covered, until ready to use.

When the greens come to a simmer, cook until the colors begin to darken and fade, 5 minutes or less. Transfer half of the greens and liquid to a food processor and puree until the greens are finely minced. Transfer to a bowl and repeat with the remaining greens and liquid.

Wipe out the pot and return the puree to it. Bring back to a simmer, then stir in the rice and vinegar. Season with a generous grinding of pepper. Taste for salt and vinegar.

Ladle the soup into heated shallow bowls and garnish with a generous grating of cheese. Serve.

Tart of Garlicky Greens and Black Olives

The slightly bitter greens turn nearly sweet during baking, and that sweetness is set off by the salty olives and piquant ricotta salata. If you can't find ricotta salata, you can use small chunks of a good brand of feta.

6 SERVINGS

- 2 tablespoons olive oil
- 3 cloves garlic, minced
- ½ pound mixed leafy greens without stems (mustard, kale, collard, beet, turnip and chard), finely chopped
- 1 15-ounce container ricotta
- ¼ cup chopped, pitted brined black olives, such as Kalamata
    Salt
- 2 large eggs
    Pastry for a 9-inch tart (see page 370)
- 2 ounces ricotta salata

Heat the oven to 400 degrees. Heat the olive oil and garlic in a soup pot over medium heat until the garlic softens, about 3 minutes. Add the greens and cook, stirring, until they soften, about 5 minutes.

In a medium bowl, beat together the ricotta, cooked greens, chopped olives and ¾ teaspoon salt. Taste the mixture; it should be highly seasoned. Add a little more salt, if necessary. Beat in the eggs.

Line a 9-inch tart pan with the pastry. Pour the filling into the pastry and shave the ricotta salata over the top. Place the tart pan on a baking sheet and bake until the center is no longer moist, 30 to 40 minutes. If you use red chard leaves, don't be fooled by what may look at first glance like quick browning; it is the color of the chard leaking through.

Let cool slightly to set up before serving.

Greens with Spicy Lemon-Cumin Oil

Use a mixture of greens to get the most complex flavor and blanch them quickly to keep their slightly bitter pepperiness intact. The spiced oil is easy to make and can be used with many other dishes.

4 SERVINGS

- 1 **pound mixed leafy greens without stems (mustard, kale, collard, beet, turnip and chard)**
- 1 **tablespoon olive oil**
- 1 **garlic clove, peeled**
- 4 **tablespoons Spicy Lemon-Cumin Oil (recipe follows)**
- 1 **tablespoon pine nuts, toasted**
  Salt

Blanch the greens, one variety at a time, in rapidly boiling, generously salted water until they are tender and bright green, 2 to 7 minutes, depending on the variety. When all the greens have been blanched, combine them into one mess. Let cool a bit. Pick up a large clump and squeeze it in your hand to wring out as much excess liquid as you can. Chop the greens finely. Repeat until all the greens are chopped.

Warm the olive oil and garlic clove in a large skillet over medium-low heat until the garlic becomes fragrant, 3 to 5 minutes. Add the greens and cook, stirring, until no more liquid remains in the skillet, 3 to 5 minutes.

Add the lemon-cumin oil 1 tablespoon at a time and cook, stirring, until each tablespoon is completely absorbed. Continue cooking until the greens are very soft, about 5 minutes. Stir in the toasted pine nuts and continue cooking for 5 more minutes. Season with salt to taste and serve immediately.

Spicy Lemon-Cumin Oil

MAKES ABOUT ½ CUP

½ cup olive oil
Zest of 1 lemon
¾ teaspoon cumin seeds
½ teaspoon red pepper flakes

Warm the olive oil, lemon zest, cumin seeds and red pepper flakes in a small saucepan over low heat until the lemon zest curls and sizzles, about 10 minutes. Remove from the heat and let steep for at least 1 hour. Strain the oil. (It will keep, refrigerated in a tightly covered jar, for 2 to 3 days.)

Potatoes

Potatoes are the essence of bland stolidity. They adapt to almost any cooking technique you can imagine and are happy being paired with almost any ingredient. Accounting for almost one third by weight of all the vegetables eaten in America and available year-round, they are so common we might well take them for granted. But that's all a facade. Dig just below the surface, and you'll find that these seemingly placid vegetables are among the strangest members of the plant kingdom.

Potatoes are members of the Solanaceae family, relatives of tomatoes, eggplants, chiles and the (really) deadly nightshade. They are tubers, which are part of the root system. But whereas true roots reach out through the soil to collect nutrients, tubers swell up to store the food the plant has collected. Tubers reproduce asexually. As every elementary school science student has learned, if you cut a potato into pieces and sow them in the ground, each piece will grow a plant exactly like the one you started with (they are true clones). Potatoes can also reproduce sexually — they have pretty white flowers that when pollinated will develop a fruit that looks like a teeny, very seedy tomato. But successful pollination occurs only very rarely, and the results are genetically haphazard — plant a potato seed, and who knows what might happen.

Potatoes are extremely demanding about their growing conditions, too. They need rich earth and are among the most nutrient-hungry of all food plants. The soil must be deep and fine; any clods will deform the spuds, so farmers are careful not to walk where potatoes are growing for fear of compacting the dirt. Potatoes also need a lot of water — about an inch per week is perfect — but the land must be well drained; soggy earth encourages rot and all kinds of fungal nastiness. Potatoes are susceptible to an amazing array of pests and diseases: early and late blights, potato beetles, mosaic virus, leafhoppers, aphids and various wilts and rots that can destroy entire fields within days. It was late blight, *Phytophthora infestans*, that devastated the Irish potato harvest in the late 1840s, killing more than a million people and creating a wholesale redistribution of population around the world. (A new, fungicide-resistant mutation of *P. infestans* has cropped up in several American growing areas.)

To help fight off this onslaught of pests and diseases, potatoes have developed their own system of self-defense. When the plant feels weak and threatened, it begins to pump up production of very potent poisons called glycoalkaloids. These are always present at low levels in the tuber and at very high levels in the leaves, flowers and stems — the most vulnerable parts of the plant. If the temperature gets too cool, if the ground gets too wet, if there aren't enough nutrients available, if pests or diseases attack the plant or if the tuber is bruised or damaged in any way, the plant begins to step up its production of glycoalkaloids.

Even after harvest, exposure to light or heat can spur glycoalkaloid production. That's why some people warn against eating potatoes that have begun to turn green on the skin. The color is caused by the development of chlorophyll rather than by the poisons themselves, but the two usually occur simultaneously. Sprouting potatoes are much higher in glycoalkaloids. Extremely rare cases of "potato poisoning" do occur, almost always the result of people eating potatoes that are spoiled or drinking tea made from potato leaves.

In most potatoes, the concentrations of glycoalkaloids are extremely low, less than 1,000 parts per million. Since the glycoalkaloids are concentrated just under the skin, you can eliminate most of them by peeling deeply. In fact, in small doses these poisons are one of the things that make potatoes so delicious, contributing a slight bitterness that nicely complements their natural sweetness and earthiness.

Potatoes grow in several stages. First, seed potatoes (pieces of potatoes) are planted. They produce a green plant above the ground. When the plant reaches a certain stage of maturity, rather than produce more foliage, it begins to pump nutrients below the ground to the tubers. Since the potato is actually a perennial (although it's farmed as an annual), the tuber stores the nutrients in the form of starches and sugars so that the plant can last the winter. Gradually, the tubers swell and grow, increasing in size. One potato plant can produce as many as a dozen potatoes, with a wide range of sizes. When the tubers have absorbed as many nutrients as they can hold, the green plant begins to wilt and die. This "lay down" signals the farmer that it is time to begin the harvest.

What happens after the harvest is almost as important as what occurs before. Although a very few potatoes are sold immediately after digging, most go into storage. Those that go straight to market are true "new" potatoes, no matter the variety or size. (The sign of a real new potato is a peel that rubs off with a thumbnail.) These potatoes are usually very moist, with a complex flavor due in no small part to their being naturally higher in glycoalkaloids. The potatoes that go to curing rooms are stored at between 60 and 70 degrees with high humidity for ten days to two weeks.

Curing helps heal whatever cuts or bruises the potato might have suffered during harvest and encourages the formation of a tough skin that will protect it during storage and shipping. Gradually, the temperature in the storage room is lowered. Although all potatoes are at their best fairly soon after harvest, they can be stored for up to ten months if the conditions are carefully maintained — between 35 and 40 degrees and high humidity.

Temperature is critical. Potatoes destined for the fresh market are kept at the bottom end of that range. Those that will be fried for chips must be held at the warmer end, or even higher. That's because chilled potatoes begin to convert starch into sugar, which will scorch when fried. The same thing can happen at home. Potatoes that have been refrigerated can taste downright sweet. When potatoes do sweeten in the refrigerator, a couple of weeks of "tempering" at slightly higher temperatures can restore some of the flavor balance.

To avoid even partial repeats of disasters such as the Irish potato famine, potato breeders are constantly working to develop new varieties with increased resistance to the spud's many enemies. The earliest European potatoes were based on varieties that originated in the Peruvian and Bolivian Andes. As the blight showed, these were very susceptible to disease. In the late nineteenth century, breeders began experimenting with Chilean varieties, which turned out to be much more resistant. This work is ongoing, with breeders today focusing on wild Mexican varieties. Russet Burbank — the gold standard for baking potatoes — traces its family tree back to the late-nineteenth-century California plant breeder Luther Burbank. Originally bred by Burbank as a high-moisture, smooth-skinned variety, the potato we know today was an accident — a genetic mutation found and then popularized by Colorado potato farmer Lou Sweet beginning around 1910.

Whereas the first half of the twentieth century was devoted to boiling, baking and mashing potatoes, the second half was all about chips and fries. Early on, the commercial production of processed potatoes was so limited that government statisticians didn't even include them as a separate category until the 1960s. By 1970 fresh and processed potatoes were grown in roughly equal amounts, and then fast food took off. In 1999 Americans ate 63 pounds of frozen potatoes per person (mostly French fries), compared to 50 pounds of fresh.

Although we are not eating as many fresh potatoes as we did in 1950, we are certainly eating more interesting ones. Today careful shoppers can choose among hundreds of varieties — including French fingerlings and German Butterballs, Red La Sodas, Purple Peruvians and Yukon Golds — with more being added all the time.

How do you decide which ones to use in which ways? The answer comes down to how much starch a potato contains — the higher the amount, the drier and fluffier the texture of the cooked potato. At the very top end of the scale are the russet, or baking, potatoes, such as Russet Burbank. When they are cooked, their starch cells swell and separate. At the opposite end of the scale are dense, moist potatoes such as the Red La Soda. The starch cells in these potatoes tend to stick together, even after cooking, giving their flesh a waxy texture that is perfect for dishes such as soups and potato salads, where you want the chunks of potato to remain intact and not fall apart. In the great middle are potatoes that are lumped under the nearly useless term "all-purpose." In truth, these potatoes demand more careful handling than those at either extreme. When baked in the traditional way, they don't have the creamy texture of true baking potatoes. They are better roasted, because they'll hold their shape and form a nice crust on their cut surfaces. Although you can use them in potato salads, you must stir very carefully to avoid turning them into mush. They are perfect for grating into potato pancakes or hash browns and are unbeatable in gratins.

Which kind of potato you choose for mashing depends on what kind of mashed potatoes you like. For a very French, very elegant puree, one that is lighter than air but rich with butter, use a baking potato. Their light and fluffy cooked texture easily forms a silky smooth liaison with the fat. People who prefer their mashed potatoes on the chunky side should use all-purpose potatoes, since their closely packed starch cells will hold together even after you beat in butter and cream.

Care is required in mashing potatoes. The starch cells in the cooked flesh are delicate; if handled roughly, they will break, and you will wind up with something akin to wallpaper paste. A potato ricer or a food mill will give the finest texture; use a potato masher or a big fork if you like your potatoes a little chunkier. Use a folding motion and a spatula or whisk to beat in the butter and hot cream or milk. The amount of butter and whether or not you choose cream or milk is up to you. The famed French chef Joël Robuchon takes the prize for the richest mashed potatoes — nearly half spuds and half butter. Escoffier, on the other hand, used only about 10 percent butter. Let your taste be your guide.

WHERE THEY'RE GROWN: Potatoes are grown all over the United States and in just about every season, although the fall harvest centered in the Pacific Northwest dominates. Winter potatoes are grown in California and Florida, and spring potatoes are harvested primarily in Florida, North Carolina and California.

HOW TO CHOOSE: Potatoes should be firm and well filled out. Avoid any that are sprouting or show signs of greening — and avoid stores that sell them. Avoid potatoes that show nicks and cuts; that's where decay starts. Don't worry about a little dirt — in fact, the best potatoes often have a little earth sticking to their skins.

HOW TO STORE: Keep potatoes in a cool, dark place. Warm temperatures encourage sprouting and shriveling. Avoid direct sunlight, which encourages the development of substances in the potato that can make you sick. Don't keep potatoes in the refrigerator, or they will turn sweet.

HOW TO PREPARE: To peel or not to peel? In most cases, there is no right answer. Nutrients are lost in peeling, but they are of negligible value. If a potato is greening, it should be peeled deeply. Other than that, it's up to your taste. Many potatoes have colorful peels, so it's good to leave them on, especially if you're steaming them or adding them to a clear soup. Scrub unpeeled potatoes with a stiff brush to remove any soil.

ONE SIMPLE DISH: Buy an assortment of waxy potatoes of various shapes and colors, preferably new ones. Cut them so they are of roughly similar size. Steam them just until tender, about 15 minutes. While the potatoes are cooking, beat together some room-temperature butter, minced shallots and chopped herbs of your choosing in a bowl. As soon as the potatoes are done, drain, add them to the bowl and stir gently with a spatula. The butter will emulsify with the little bit of water left on the potatoes and form a sauce that will adhere lightly. Sprinkle with fleur de sel and serve immediately.

Potato and Green Bean Salad with Green Goddess Dressing

All cooks have a few basic recipes that they turn to again and again over the course of a year. Potato and green bean salad is one of mine. I make it different ways depending on the season and my mood. It's very good dressed with just olive oil and lemon juice, but it becomes absolutely superb when bound with homemade Green Goddess. If you're familiar only with the bottled version of this dressing, you must try my recipe, which is based on the original, invented in the 1920s by the great San Francisco chef Victor Hirtzler.

6 SERVINGS

- 1½ **pounds new potatoes**
- 1 **teaspoon salt**
  Green Goddess Dressing (recipe follows)
- ¾ **pound green beans**

Cut the potatoes into roughly equal pieces about the size of a small walnut. Place them in a steamer basket over boiling water, cover and cook until tender, about 15 minutes. Transfer from the steamer to a large bowl and immediately season with the salt and 2 tablespoons of the dressing. Stir well to combine and set aside.

Trim the stem ends of the beans and, if necessary, remove any strings. Steam the beans as you did the potatoes. It will take about 7 minutes for them to become bright green and just barely tender.

Add the beans to the potatoes along with just enough dressing to coat lightly, about 2 tablespoons. (The remaining dressing can be kept in a tightly covered container in the refrigerator for up to 1 week.) Serve at room temperature. If you refrigerate the salad, bring it to room temperature before serving.

Green Goddess Dressing

1½ cups mayonnaise
4 salted anchovy fillets
2 green onions, green parts only
2 tablespoons chopped fresh tarragon
2 tablespoons snipped fresh chives
1½ tablespoons chopped fresh parsley
2 tablespoons tarragon vinegar

Combine the mayonnaise, anchovies, green onions, tarragon, chives, parsley and vinegar in a food processor or blender and process to a rough puree. Or grind with a mortar and pestle. Refrigerate, tightly covered, until ready to use.

Gratin of Potatoes, Leeks and Mushrooms

Gratins are basically nothing more than vegetables and cream baked together in a slow oven until they fuse. The vegetables soak up the liquid to the point that they become meltingly tender. The cream ennobles the essence of each vegetable. The crusty layer on top combines the best of both ingredients and gilds the dish with a lovely brown. No wonder gratins are so beloved.

But gratin lovers are a contentious bunch, willing to argue almost any point: the shape of the pan, the type of potato, the mixture of cream and milk. That's odd, considering that the finished dish is such a paean to simple comfort and that it always seems to turn out perfectly, no matter what you do. I tested this dish with baking and boiling potatoes side by side and couldn't tell the difference.

6 SERVINGS

- 1 **tablespoon butter**
- 1 **leek, white part only, chopped (about ¾ cup)**
- ¼ **pound mushrooms (any type), sliced**
- 1 **shallot, minced**
- **Salt**
- 1½ **pounds potatoes, peeled**
- 1½ **ounces Gruyère, grated (about ¼ cup)**
- 1¼ **cups milk**
- ⅔ **cup heavy cream**

Heat the oven to 400 degrees. Generously butter a heavy baking dish that is roughly 8 by 10 inches.

Melt the 1 tablespoon butter in a skillet over medium heat. Add the leek and cook until softened, about 5 minutes. Add the mushrooms and shallot and reduce the heat to low. Cook until the mushrooms begin to darken and give off their moisture, about another 5 minutes. Season with salt to taste.

Slice the potatoes very thinly and pat them dry with a kitchen towel. Arrange half of the potatoes in a rough layer in the baking dish, then distribute the leek and mushroom mixture over the top. Scatter half of the grated cheese over that and season with salt to taste. Top with the remaining potatoes in as solid a layer as possible. Scatter the remaining cheese over the top.

Bring the milk just to a simmer in a small saucepan and pour it over the potatoes. You should just barely be able to see the milk under the top layer. Pour the cream over the top and bake for 30 minutes. The milk should be puffed and bubbling and beginning to brown, and the potatoes should be starting to become tender. Bake for 25 to 30 minutes more, or until the top is thoroughly browned, the milk bubbles have subsided and the gratin is a compact mass. Remove from the oven and serve.

Root Vegetables

● ● ● ● ● ● ● ● ● ● ● ● ● ● ● ● ●

Perhaps no part of a plant is as complex or requires such special handling as the roots. But few other parts are so rewarding — and just at the times we need them most. Roots serve a variety of functions. They provide a firm base from which a plant grows upward. They probe downward in search of moisture, collect it and begin the process of sending it up to the leaves, where it's most needed. In turn, they store the nutrition that the leaves produce.

Although the group that we call "root vegetables" is wide and varied, it's important to remember that, botanically speaking, not everything that grows underground is a root. True roots include carrots, parsnips and sweet potatoes. But regular starchy potatoes, which grow even deeper under the surface, are tubers. Other vegetables that we consider roots are more technically the junction between the roots and the stem, growing partly aboveground. These include beets, turnips and radishes.

Structurally, though, all of these varied parts are roughly similar. They are all covered with a rough, corky skin, which serves to protect the plant's interior from damage and retain moisture. That these peels are so spectacularly successful at these jobs is why we can eat root vegetables all winter (although they are usually no longer stored in root cellars). We usually peel this skin before we eat the vegetable. In many cases, because the peel is made of a different material, it is so loosely attached to the root that after cooking, it will slip free of its own accord. In the center of the root is the section that transports water from the earth to the plant's aboveground parts. This is called the vascular system. It is frequently tough and woody, and in the cases of carrots and parsnips, careful cooks may remove it as well. In between the peel and the vascular system is the part that we really crave. This is the storage system, which is not only tenderer than the vascular system but also where the plant stores the sugars and starches that it will need to grow. (The arrangement is not always so simple. Beets alternate vascular and storage layers, in some cases even colored with different pigments, producing a striped "bull's-eye" effect.) Root vegetables can be astonishingly sweet. The sugar beet, a cousin of the familiar red beet, is processed to make sugar.

The end result of all of this heavy construction is something that usually looks absolutely unpromising in the raw ("Where did you dig *that* up?") but that turns incomparably sweet, flavorful and scented with a little careful cooking. Just when we need these vegetables most, in the bleakest part of the winter, they give us the sustenance to carry on until spring. Their texture and sweetness come from a combination of starches and sugars. To the plant, starches represent food that has been stored for future use, while sugars can be immediately converted to energy. Starches are chemical compounds that resemble tough little pellets when raw. After they are heated in combination with a liquid, they soften. (In this way, what happens when you cook a beet is not all that different from what happens when you make a simple white sauce.)

Sugars are closely related to starches. (Simple and complex carbohydrates, remember?) In fact, enzymes produced by the plant can convert starches (stored food) to sugars (usable food) when doing so is necessary for the plant's survival. This is why parsnips are almost always sweeter when harvested after a hard frost: the plant, feeling threatened by cold weather, has started converting its stored food to food that it can use immediately. By contrast, a plant such as the carrot, which hails from a more moderate climate, has a higher ratio of sugar to starch and is sweeter right off the bat.

Properly selected and prepared, all root vegetables turn sweet. But this is only part of their appeal. What makes root vegetables so fascinating is the diversity of secondary attributes they possess. In general these fall into two main categories, the sharp and the colorful. The first typifies the flavors of vegetables such as radishes, kohlrabi, rutabagas, turnips and horseradish and comes from varying concentrations of a mustardy sulfurous compound that in nature functions as a defense mechanism. As with onions, cooking tames the heat by altering the enzyme that helps create it, as does pickling in an acid such as vinegar. Pungent as these roots can be in flavor, they are, to varying extents, bland in appearance (although radishes have thin skins that can be colorful).

Other roots, while lacking inner fire, are among the most vibrantly colored members of the vegetable kingdom. The most obvious example is the beet, which, depending on the variety, is colored anything from blood red to golden orange or even pure white. This color comes from the pigment betalain — or its absence. (Betalain also gives bougainvillea its distinctively fiery colors.) Unlike most plant pigments, betalain is water-soluble and will leak readily — as anyone who has cut up beets on a wooden chopping block will sadly attest. For this reason, beets are usually treated differently than other vegetables — cooked whole before they're peeled. Once cooked, the corky peel will slip right off (although you still have to be careful of staining).

We usually think of carrots as being bright orange, colored with a pigment called, appropriately, carotene. But they, too, come in a wide variety of hues. In fact, the familiar orange carrot probably dates back only to the seventeenth century. Some people believe that the original carrot was a purplish red color nearly identical to that of the beet. These carrots are colored by the pigment lycopene, which is also found in sweet potatoes, as well as in tomatoes, watermelons and pink grapefruits. Other carrots are colored by anthocyanins, pigments that give the reddish color to everything from berries and grapes to autumn leaves. Because all these pigments are so much more colorfast than betalain, the dull corky skin of carrots can be peeled before cooking.

Root vegetables adapt well to both moist- and dry-heat cooking methods. Moist heat, such as that from boiling or steaming, softens the vegetables' starch and cellulose more quickly and keeps colors brighter and flavors purer and more direct. Dry-heat cooking (primarily roasting) takes longer. The colors tend to be darker and not quite so fresh, and the flavors developed are more complex.

Because of the higher temperatures attained during roasting, the root's sugars will begin to caramelize, and you will begin to get some of the flavors from browning associated with the Maillard reaction. Most roots can be nicely roasted simply slicked with a little oil. Cooking will be a little faster, and the result a little moister, if the vegetables are wrapped in aluminum foil first. This is absolutely necessary for beets, if for no other reason than to keep them from bleeding all over the kitchen.

Beets

• • • • •

WHERE THEY'RE GROWN: Beets are harvested primarily in Wisconsin, New York, Oregon and Texas.

HOW TO CHOOSE: Beets are usually sold with their tops on. Inspect the greens for freshness. Also, avoid any beets with cracks or soft spots or that have a lot of hairy secondary roots.

HOW TO STORE: Store beets in a plastic bag in the refrigerator.

HOW TO PREPARE: Beets will bleed pigment all over everything. Don't peel them until after they've been cooked; they will keep their color better. Also, leave about an inch of tops on and don't break off the bottom root for the same reason.

ONE SIMPLE DISH: Wrap beets in aluminum foil and roast at 400 degrees until they are soft enough to be pierced with a knife, about 1 hour. Peel them, quarter them and dress them with olive oil and red wine vinegar. Season with salt to taste and finish with a generous grinding of pepper.

Carrots and Parsnips

· · · · · · · · · · · · · · · · ·

WHERE THEY'RE GROWN: California grows more than two thirds of the carrots that are harvested in the United States. California also leads in parsnip production.

HOW TO CHOOSE: Carrots and parsnips with the tops on are great because you can tell from the greens how recently they were harvested. But sometimes very fresh vegetables are sold without the tops. The best trick is to pay attention to the intensity of the color. The pigments fade with time, so the deeper orange a carrot is, the more likely that it's fresh. Avoid vegetables with splits and cracks and those with lots of little hairy roots; they are too old.

HOW TO STORE: Store carrots and parsnips in a plastic bag in the refrigerator. Remove the tops if you're not going to be eating them right away, as the greens will pull moisture from the roots.

HOW TO PREPARE: Carrots and parsnips should almost always be peeled. The peel is corky and cooks to a different consistency than the core. Also, if they have a large, pale center, quarter them lengthwise and cut it out. It will be woody and flavorless.

ONE SIMPLE DISH: Carrots and parsnips are delicious braised. Peel them and slice them into rounds or sticks. Place them in a skillet with enough water just to cover the bottom of the pan and a knob of butter (and maybe a hint of minced garlic or shallots). Cook, covered, over medium heat until almost tender, then remove the lid, increase the heat to high and cook until the liquid evaporates.

Radishes

• • • • • • • •

WHERE THEY'RE GROWN: Florida grows more than half the radishes produced in the United States, followed by California, Michigan and Ohio.

HOW TO CHOOSE: Radishes are almost always sold with their tops. That's the first thing to check — they should be bright green and not at all wilted. The roots should be brightly colored and free from cracks and nicks. Give them a squeeze: if they're not hard, they could have a soft center.

HOW TO STORE: Store radishes in a plastic bag in the refrigerator, removing the tops if you're not going to use the radishes right away.

HOW TO PREPARE: Radishes need only a thorough washing to be ready to eat. Don't skip this step, though. Because they're grown in fine, sandy soil, grit can show up even where it's not obvious.

ONE SIMPLE DISH: Wash radishes in ice water, then arrange them on a plate around a crock of softened butter and a bowl of coarse salt. To eat, rub them in the butter, then dip them in the salt.

Sweet Potatoes

• • • • • • • • • • • • •

WHERE THEY'RE GROWN: North Carolina is the leading state for sweet potatoes, followed by California, Louisiana and Mississippi.

HOW TO CHOOSE: There are two kinds of sweet potatoes in markets. One is pale orange and starchy; the other is dark orange, a little sweeter and very moist. Sweet potatoes are frequently called "yams," although they are completely different from true yams, which hail from West Africa. Use the dark orange ones when you want the sweet potato to hold together in a dish. Use the pale ones when you want a dish to be light and fluffy.

HOW TO STORE: Sweet potatoes will last for a couple of weeks at cool room temperature, but refrigerating does them no harm.

HOW TO PREPARE: Sweet potatoes need only to be peeled before cooking. If you are going to roast them, even that isn't necessary — you can spoon the pulp from the skin after they are cooked.

ONE SIMPLE DISH: Pierce sweet potatoes with a fork in several places. Place them on a jelly-roll pan and bake at 450 degrees until they are quite soft, 30–40 minutes. (Piercing keeps them from exploding, and the jelly-roll pan will catch the sugar syrup they exude.) Spoon the pulp from the skin and beat it with a couple of tablespoons of butter, a little salt and a grating of nutmeg.

Turnips and Rutabagas

WHERE THEY'RE GROWN: Farmers in two different areas of the United States grow turnips. Tops are farmed in the mid-South, primarily in North Carolina and Georgia. Bottoms are grown in California. Illinois grows both tops and bottoms. Rutabagas, which are a cross between turnips and a kind of cabbage, are too sparsely grown to be tracked statistically.

HOW TO CHOOSE: Choose roots that are free from nicks and scars. Check the top of each turnip or rutabaga, where the greens once were. As the roots sit, they will continue to sprout new greens, which the produce manager will trim. The more the tops have been trimmed, the older the root probably is. In Japanese markets you can find small white turnips that have the tops attached. These are milder and sweeter than other turnips.

HOW TO STORE: Store turnips and rutabagas in a plastic bag in the refrigerator.

HOW TO PREPARE: Turnips and rutabagas should be peeled, except for the small Japanese varieties, which have such a thin skin.

ONE SIMPLE DISH: Braise turnips or rutabagas just as you would carrots (see page 353).

Cream of Parsnip Soup

This is a somewhat plainer version of a recipe by the great San Francisco chef Jeremiah Tower. He garnishes his version with shaved white truffles. It's also really, really good with sour cream.

4 SERVINGS

- 1 **pound parsnips**
- 1 **tablespoon butter**
- 1 **onion, chopped**
- 1 **medium boiling potato, peeled and diced**
  Salt
  About 3⅓ cups water
- 1 **sprig tarragon**
- 1 **sprig parsley**
- ¼ **cup sour cream**

Working lightly with a vegetable peeler, peel the parsnips, then cut off the bottoms and tops. Continuing to use the vegetable peeler, cut away and save the rest of each parsnip down to its woody core, catching the thin slices in a wide pot. The color of the vegetable will change from creamy white to ivory when you get to the core. Discard the core.

Add the butter, onion, potato and 1 teaspoon salt to the pot along with ⅓ cup water. Place the pot over low heat, cover it tightly and cook slowly, "sweating" the vegetables until they begin to become tender, about 15 minutes. Stir from time to time to keep the vegetables from sticking and scorching. If necessary, add a little more water.

Add the tarragon and parsley and continue to sweat for another 5 minutes. Add 3 cups water, increase the heat to medium and cook, uncovered, until the vegetables are completely tender, about 10 minutes.

Discard the tarragon and parsley sprigs and, using a slotted spoon, transfer as much as you can of the solids from the pot to a blender. With the lid of the blender removed, pulse to chop the vegetables. If necessary, add a little water. Once the vegetables are chopped, blend on the lowest speed and gradually work your way up to the highest. At first the vegetables will jump up the sides, but then they'll subside and remain at much the same level no matter the speed of the blender. With the motor running, add the rest of the liquid and any vegetables left over in the pot and puree until completely smooth.

Wipe out the pot to remove any bits of vegetables, then pour the pureed soup back into it. Heat through over low heat. Taste for salt.

Beat the sour cream with a spoon to soften it. Divide the soup among four warmed soup bowls, drizzle in a bit of sour cream in a decorative pattern and serve.

Turnip and Potato Gratin

Getting the maximum mileage from two quite humble ingredients, this gratin combines the earthiness of the potato with the sweetness of the turnip, and it's a breeze to make for a crowd. An Irish friend insists that I make it every Christmas because it reminds him of "neeps and tatties."

8 SERVINGS

- 1 garlic clove, peeled
- 6 turnips, peeled
- 2 large boiling potatoes, peeled
- 2 teaspoons salt
- 1½ cups heavy cream
- 3 ounces Gruyère or Comte, grated or sliced

Heat the oven to 450 degrees. Rub the garlic clove all over the inside of a heavy gratin dish, then butter the dish well and set aside. Discard the garlic.

Slice the turnips and potatoes as thinly as you can, ideally using a mandoline or Japanese slicing tool. Toss the vegetables with the salt and place them in rough layers in the gratin dish. Don't worry about arranging them; you'll be stirring them later. Bake until softened, 20 to 30 minutes. Stir with a spatula every 10 minutes, making sure the bottom layer doesn't scorch.

Pour the cream over the potatoes; it should come just to the top layer without covering it. Distribute the cheese over the top and bake until the cream thickens and the top is browned, about 30 minutes. Serve hot.

Sweet Potato–Prosciutto Soufflé

This recipe is best with the pale gold sweet potatoes that are frequently labeled "yams." They are a little less sweet and much drier and starchier than the dark orange ones. These soufflés are not the delicate little darlings you may have feared. They can even be made in advance and frozen, and they'll still puff up.

6 SERVINGS

- 1 **medium to large sweet potato (see headnote)**
- 3 **tablespoons ricotta**
- 2 **large egg yolks**
- ¾ **teaspoon salt**
- 3 **slices prosciutto**
- 5 **large egg whites**
- ⅛ **teaspoon cream of tartar**

Heat the oven to 375 degrees. Put the sweet potato on a baking sheet and bake until it is easily pierced to the center with a knife, about 1 hour. Set aside until cool enough to peel. Peel the potato and puree the flesh in a food processor. Reserve ⅔ cup of the puree. Leave the oven on. (You can save the rest for up to 1 week, tightly covered and refrigerated.)

Generously butter six straight-sided ½-cup ramekins or individual soufflé molds.

Return the sweet potato puree to the food processor and add the ricotta, egg yolks and salt. Process to a smooth puree.

Roll the prosciutto slices into a tight bundle and slice crosswise into very thin strips. Scatter them with your fingertips to keep them from sticking together.

Beat the egg whites in a large, clean bowl until frothy. Add the cream of tartar and continue beating until stiff peaks form. Do not overbeat.

Add about ½ cup of the sweet potato–ricotta puree to the egg whites and gently fold in. The best way to do this is with a handheld balloon whisk or a spatula. In either case, cut straight down through the puree and the egg whites, scrape the bottom and lift the egg whites over the puree. Turn the bowl a quarter turn and repeat. Keep doing this until the puree is fully incorporated, 3 or 4 times.

Add the remaining sweet potato–ricotta puree and fold it in the same way. When you're done, the puree should be evenly distributed, but there may be small patches of egg whites remaining. Do not overmix, or the egg whites will lose too much volume.

Scatter the prosciutto strips over the top and very gently fold them in. Do not worry too much about distributing them evenly.

Using a large soupspoon, evenly divide the soufflé mixture among the buttered ramekins. Depending on how well the egg whites were beaten, you may have some of the mixture left over. Either bake in another ramekin or discard.

(The recipe can be made to this point, covered tightly with plastic wrap and refrigerated for up to 2 hours or frozen for up to 2 weeks. The soufflés can be baked straight from the refrigerator. If frozen, remove the plastic wrap immediately and let stand at room temperature for 30 minutes before baking.)

Put a jelly-roll pan in the oven and arrange the ramekins on it. Bake until the tops are puffed and dark brown in spots, 25 to 30 minutes. The surest sign of doneness is when the soufflés become extremely fragrant. Do not overbake, or the centers will be dry.

Remove from the oven and serve immediately.

RELIABLE SOUFFLÉS

A good soufflé is a transcendent bit of cooking, delivering a wallop of flavor on a breathy whisper. It is intense yet ethereal, profound but insubstantial. True, it has a reputation for being the diva of the food world — risky and temperamental. But in reality it's a surprisingly tough little rascal, and its rewards far outweigh the bit of extra trouble.

Did you know that a soufflé can be put together in less than 20 minutes? That you can make it ahead and freeze it until just before dinner? Now I've got your attention.

The magic of the soufflé is how its fairly normal-looking batter puffs and fills with air while baking. Like so many other bits of kitchen wizardry, that loft comes thanks to the egg — specifically, the egg white, which is a combination of water and protein and not much else. When egg whites are beaten with a whisk, the proteins, which are naturally curled up in little balls, relax and unfold. As they do, they connect with other proteins and form bonds. These connected strands leave small gaps — tiny bubbles, really — filled with air. When the soufflé is heated, the air expands inside the bubbles and the soufflé puffs. The water evaporates, leaving nothing behind but the inflated thin framework of protein strands — and whatever else you add to them, of course.

No matter how interesting they may be scientifically, there's nothing very compelling flavor-wise about plain egg whites. Traditionally, the addition of flavor begins with some kind of flour-based paste, such as a thick white sauce. But there is nothing flavorful about that either. You can get the same effect — and a lot more flavor — by using other kinds of pastes: a puree of roasted sweet potatoes, for example, or a soft cheese such as ricotta or fresh goat cheese. To this base, egg yolks are usually added. They make the soufflé a little richer, but they also thin the paste enough that it won't collapse the egg whites when they're folded in.

Although the base carries the flavor, by far the trickiest part of making a soufflé is the egg whites — and they are probably the reason for much of the recipe's reputation for being difficult. They need to be beaten — but just enough. They need to be handled gently. Most cooks have some kind of electric mixer for beating egg whites, either a handheld one or a big stand mixer.

After only a couple of minutes, you'll notice that the egg whites are beginning to hold a shape: soft, billowing mounds. Keep going. Within a minute or two, you'll notice that the whites are forming something that actually looks like a peak (as opposed to a small hill). Lift the beaters from the bowl: the whites will probably form a point at first, then almost immediately collapse back into the bowl. These are called "soft peaks," and they aren't stable enough to withstand the heat of the oven. When the peaks are stiff enough that they hold their sharp points — both in the bowl and from the beaters — you're done. The beaten whites should be shiny and glossy. Roll the bowl around; there should be no loose whites in the bottom of the bowl. If you want to make certain they're done, prop a whole egg on the whites — they should be firm enough to support it.

Don't overdo the beating, though. Stiffer is not better. You can beat in so much air that the structure no longer contains bubbles, and you'll be left with a clumpy, grainy mess that looks like Styrofoam. If this happens, you have to start over.

When beating egg whites, it's important to have a very clean metal bowl and beaters. Any trace of fat will interrupt the linking of the protein chains, and the bubbles won't inflate. Don't use plastic or wooden mixing bowls, which are so porous that you can never be sure they are completely free from fat.

Stiff-peaked egg whites are so incredibly stable that you can spoon the fully prepared soufflé into a ramekin, cover it tightly with plastic wrap and refrigerate it for a couple of hours with no ill effects. You can even freeze a soufflé for up to a month. Just pull it out about 30 minutes before baking to let it warm up slightly so the dish won't shatter in the oven. (If you use CorningWare or Pyrex, you don't even need to do this.) Be sure to remove the plastic wrap as soon as the soufflé comes out of the freezer. Do it later, when the egg whites have defrosted and are more delicate, and you risk an embarrassing deflation.

Lemons and Limes

· · · · · · · · · · · · · · · · · · ·

Fruits are for eating; lemons and limes are for seasoning, like a pinch of salt or a grinding of pepper — or, perhaps more to the point, a splash of vinegar. Fruits are sweet and seasonal; lemons and limes are sour to the point of puckery, and you would no more expect to find them missing from the grocery store than you would onions or garlic. But those lemons and limes we so thoughtlessly squeeze onto a slab of broiled salmon are grown, not manufactured — no matter what you might think when you see them lined up so perfectly at the grocery. In fact, they qualify as heirloom fruits. The two main varieties of each that are grown today were introduced more than one hundred years ago.

Lemons and limes are thought to have originated in the same part of Asia, but today lemons are preferred in areas where the weather is milder — most of Europe, for example — while limes are preferred where the weather is hot and humid — the Caribbean basin and Southeast Asia.

Lemons are among the most ancient of the citrus hybrids. They probably first appeared in the foothills of the Himalayas, but records indicate that they reached the Mediterranean basin as early as the first century A.D. They were among the fruits spread by the Arabs during their domination of southern Europe. Lemons reached the New World on Columbus's second voyage, in 1493.

Most of the lemons you'll find in the grocery store today are either Eurekas or Lisbons. They look pretty much the same — lemon varieties are notoriously difficult to tell apart, even for the people who grow them. The only real clues are that the skin of the Eureka is slightly ridged and a little rougher than that of the Lisbons. The best way to know which type you're buying is by the time of year. Lisbons, which originally came from Portugal to the United States in the middle of the nineteenth century, are cool-weather lemons and are harvested in the winter and spring. Eurekas were discovered in 1858 in Los Angeles, where they originally sprang from seeds of an Italian variety. They handle the heat better than Lisbons and are picked in the spring and summer, but they can hang on the tree for months after ripening and often are available even in the heart of the Lisbon harvest.

The Bearss, discovered in a Florida grove in the 1950s, is the lemon of choice in the small Florida lemon industry. Of more interest to cooks is the Meyer lemon, which isn't really a lemon at all, although it certainly looks and behaves like one. It has a sweeter flesh and a more aromatic peel than most true lemons. Discovered near Beijing, China, by a plant explorer named Frank Meyer, it was introduced in the United States in 1908. Until recently, it was believed to be a cross between a lemon and a mandarin, but DNA testing has found that it is really the offshoot of a union between a lemon and a sweet orange. It has wonderful eating qualities, but because it's so juicy and its peel is so thin and delicate, the Meyer was long regarded strictly as a backyard fruit. It became a favorite of California cooks and farmers' market growers in the 1990s and is now available nationally as a specialty.

If anything, limes are even more venerable than lemons. Their early history is clouded by a confusion of names. In some Asian and Arabic languages, the fruit is lumped in with lemons and other acidic citrus. We can be reasonably certain that the lime spread west at the same time as the lemon and by the same means — carried along by the Arab conquest and by the crusaders. The story of the modern lime begins in the sixteenth century, when Spanish and Portuguese sailors introduced it in the Caribbean basin. With the ebb and flow of colonization, many lime orchards were abandoned, and the trees were left to nature to sport and cross-pollinate as they pleased. From these Caribbean limes gone wild, two important varieties emerged: Tahiti and West Indian.

In most of the lime-loving world, the West Indian is the preferred variety. It is a small, round lime that is greenish yellow at full maturity, extremely sour, and intensely aromatic and flavorful. It is widely known in the United States as the Key lime and is found today in slightly different versions throughout India, Egypt, Morocco, Brazil and Mexico under various local names. At the turn of the twentieth century, the West Indian lime was just as important in the United States as it now is in the rest of the world. But in 1926 a massive hurricane in southern Florida wiped out the lime orchards, and today this lime grows there mainly as a backyard fruit. These limes also show up in Mexican markets, where they are labeled "Mexican limes."

The lime that replaced the West Indian in the hearts of American growers was the Tahiti, sometimes called the Persian (though, oddly, there is no trace of such a lime in Iran). The Tahiti is much larger than the West Indian and largely seedless. Although the flavor and aroma of the Tahiti are not as powerful as those of the West Indian, the tree is stronger — resistant to many of the fungi and viruses that afflict the West Indian. It is also much less cold sensitive and does not require as much heat to ripen. Prior to Hurricane Andrew — which in 1992 devastated the area around Homestead, Florida, that was the center of Florida's lime culture — more than 6,300 acres of the fruit were cultivated in the state. Ten years later, only 800 acres remained. An infestation of citrus canker in 2002 completely wiped out the lime harvest in Florida, and today all of the limes — both fresh and processed — sold in the United States are imported, most of them from Mexico.

Because lemons and limes are valued mainly for their acidity, which decreases with ripeness, they can be picked as soon as they reach minimal size and juice content, rather than at a certain stage of maturity. Fruit that is harvested when it is young lasts longer than fruit left to ripen further. The fruit is then sorted by color according to maturity and cured in a controlled atmosphere — 45 to 55 degrees for lemons and 50 to 55 degrees for limes, both at 85 to 95 percent humidity. During this process, the rind thins and becomes more intense in color, and the flesh, which is hard and dry at harvest, becomes juicier. Both lemons and limes are very cold sensitive. If they are stored below 50 degrees for too long, the rind will become pitted and splotchy, and decay will accelerate. Fruit that is harvested at an early age is prone to what those in the trade call "oil spotting" — brown stains on the surface of the rind, caused when the oil cells rupture after the fruit has been bumped and bruised.

Although most people tend to think of squeezing a lemon or lime for the juice, good cooks know that the most flavorful part of the fruit is the peel, or, to be more specific, the surface of the peel. That part is appropriately known as the "zest," where the fruit's aromatic oil glands lie. The juice of a lemon or lime may have overtones of flavor, but it is obscured by the extreme acidity. With the zest, there is very little pucker to distract from the fruit's wonderful aromatic qualities. For this reason botanists call this part of the peel the "flavedo." Just underneath it is the "albedo," the bitter part you don't want to eat. The albedo is composed of white, pithy material — the main conduit for water and nutrition when the fruit is on the tree. Culinarily, the albedo is used almost exclusively as a source of pectin, which encourages jelling.

WHERE THEY'RE GROWN: Lemons are grown mostly in California, with some coming from Arizona and a few from Florida. Limes are almost entirely imported, with most coming from Mexico.

HOW TO CHOOSE: Choose fruits that are heaviest for their size and have the strongest perfume. With Meyer lemons, look for fruit with a baby-soft, thin peel. With West Indian (Key) limes, remember that the yellower they are, the riper.

HOW TO STORE: Lemons and limes should be stored at room temperature. Chilling will cause them to spoil more rapidly.

HOW TO PREPARE: If you want lemon or lime zest and don't have a zester, use a sharp, thin-bladed paring knife to cut away just the colored part of the peel, leaving behind as much of the bitter pith as you can. Lay the peel cut side up on a cutting board, press the knife blade flat against it and slice away any traces of pith. Then you can cut the zest into thin strips.

ONE SIMPLE DISH: You can make a simple sauce for pasta by simmering heavy cream and lemon zest, then adding a little minced fresh parsley and maybe some pine nuts just before serving. Remember that the sauce should only lightly coat the pasta, not puddle in the bottom of the bowl.

Lemon and Pistachio Panna Cotta

This recipe is adapted from Los Angeles pastry chef Nancy Silverton's justly celebrated version made with bitter almonds. You can serve this as is or with sugared fruit such as strawberries.

4 SERVINGS

- 1½ **cups heavy cream**
- ¾ **cup milk**
- 3 **tablespoons sugar**
- ½ **cup chopped pistachios**
- 2 **tablespoons grated lemon zest**
- 1 **teaspoon unflavored gelatin**
- 2 **tablespoons cold water, plus more if needed**

Combine the cream, milk and sugar in a saucepan over medium heat and cook until very warm, about 5 minutes.

Combine the pistachios and lemon zest in a food processor. Pour the heated cream mixture over the top and pulse 3 or 4 times to break up the pistachios. Transfer to a covered container and refrigerate for 2 hours to infuse the flavors.

Place the gelatin in a large stainless steel bowl. Pour the water over the gelatin, adding more water if necessary to moisten all of the gelatin.

Place the bowl over a saucepan of barely simmering water and heat until the gelatin melts, about 30 seconds. Do not stir, as it will scald on the sides of the bowl. Turn off the heat and keep the gelatin warm.

Using your fingers, lightly coat four ½-cup ramekins with vegetable oil.

Bring the cream mixture to a simmer and pour it through a fine-mesh strainer over the gelatin. Whisk to combine thoroughly, scraping the bottom of the bowl to free any gelatin that may have solidified there.

Divide the mixture evenly among the ramekins and refrigerate for 6 to 8 hours.

To unmold each panna cotta, run a thin knife around the inside of the ramekin, dip the bottom in simmering water for a few seconds and invert the panna cotta onto a plate. Serve.

Lemon Curd Tart

This is an old favorite that I make several times a year. It's best made with Meyer lemons. The tart filling is my idea of the ultimate curd, buttery and lemony in perfect balance. The combination of creamy, tart curd and crisp shell makes it an ideal ending for a rich winter dinner. Make the curd while the tart shell chills and bakes.

8 SERVINGS

Tart Shell
- 1¼ cups all-purpose flour, plus more for rolling
- 1 tablespoon sugar
- Pinch salt
- 8 tablespoons (1 stick) unsalted butter, cut into small pieces
- 2–3 tablespoons ice water

Curd
- 2 large eggs
- 2 large egg yolks
- ½ cup sugar
- ¼ teaspoon salt
- Zest of 1 lemon
- ½ cup lemon juice
- 6 tablespoons (¾ stick) cold unsalted butter, cut into pieces

For the tart shell: Butter a 9-inch tart pan with a removable bottom. Set aside.

Combine the flour, sugar, salt and butter in a food processor or large bowl and cut them together until the mixture resembles coarse cornmeal. Add the water 1 tablespoon at a time, stirring constantly or processing until the mixture just begins to come together.

Remove the dough from the bowl and knead lightly and briefly to make a smooth mass. Wrap it in plastic wrap and refrigerate for at least 30 minutes.

On a well-floured work surface, roll the dough into a circle about 11 inches in diameter. Roll the dough back onto the rolling pin and transfer it to the tart pan. Unroll the dough and gently press it into the pan. Trim the excess dough to 1 inch from the pan edge and tuck the extra dough between the pan and the dough rim to make a sturdier, taller edge. Refrigerate for 30 minutes.

Heat the oven to 425 degrees. Prick the crust with a fork. Line the crust with a sheet of aluminum foil and fill it with rice, dried beans or pie weights. Bake for 10 minutes. Remove the foil and beans and bake until the crust is golden brown and firm, about another 15 minutes. Remove from the oven and let cool to room temperature.

For the curd: Beat the eggs, yolks, sugar and salt in a small saucepan until smooth and light colored.

Add the lemon zest, lemon juice and butter and cook over medium heat, stirring constantly, until the butter melts, about 2 minutes.

Reduce the heat to medium-low and continue cooking and stirring for about 5 minutes, or until the curd is thick enough to coat the back of a spoon and your finger leaves a definite track when you draw it across the spoon. The curd should resemble a thick hollandaise. Pour it through a fine-mesh strainer into a chilled bowl.

Spoon the curd into the prepared crust and smooth the top with the back of a spoon. Refrigerate for at least 1 hour to set the lemon curd. Serve cold.

CITRUS CURD

Lemon curd is a paradox in a saucepan: it's rich and creamy; it's tart and fresh. Those are the characteristics that make it delicious; they're also the characteristics that make it seem impossible to prepare. Think about it: What happens when you pour lemon juice into milk? How would you make that taste good? The answer lies in a particularly basic and very useful bit of food chemistry.

Citrus curds are creamy despite having little or no cream in them. Instead, the luscious texture is supplied by cooked beaten eggs. But citrus curds also are smooth, not like scrambled eggs at all. How does this work?

First, you need to understand a little egg chemistry. Eggs are full of strands of protein that at room temperature are tightly curled and separate. As they heat up, the protein strands relax and unfold. As they unfold, they bump into other protein strands and link up, capturing the liquid that is present in the egg. This is called coagulation. Typically, with pure egg, the white begins to set between 145 and 150 degrees and is firm at 160 degrees. By 180 degrees the protein strands have tightened to the point that all the liquid is wrung out. (Egg proteins are not alone in this behavior; that's what happens when you overcook a chicken breast, too.)

That 145-to-160-degree window is pretty hard to hit, especially over a live flame, which is precisely why perfectly scrambled eggs are such a miracle. Add sugar, however, and an interesting thing happens. The window opens wider. The sugar isolates the protein strands, moving them farther from each other and keeping them apart longer. This raises the temperature at which they coagulate. Cooks have been taking advantage of this chemical reaction for centuries. The phenomenon is the secret behind both cooked custards and zabaglione, the Italian dessert consisting of egg yolks, sugar and, traditionally, Marsala that is beaten and cooked until it is a slightly foamy cream.

In fact, if you take zabaglione, replace the liqueur with lemon juice and add butter, you have lemon curd.

But there are lemon curds and there are lemon curds. To come up with the version I liked best, I first analyzed a dozen recipes from different cookbooks. Although the techniques were basically the same, the proportions of ingredients varied greatly. Some recipes, for instance, called for whole eggs, some for yolks and some for a combination of the two. I made curds with pure yolks and with an equal volume of whole eggs and compared them. Both methods worked equally well, but there were major differences in the outcomes. Made with yolks, the curd was stiffer and the flavor richer and more custardy. Made with whole eggs, the curd was lighter-colored, the set was softer, and the flavor was more intensely fruity. I ended up compromising, using 2 whole eggs and 2 egg yolks to give the curd a slightly smoother flavor and slightly firmer set.

The amount of sugar in the recipes varied even more, ranging from ½ cup to 1 cup. A recipe I made with ½ cup of sugar was well balanced. I tried ¾ cup and found that it was too simply sweet, particularly in the aftertaste. I never added as much as 1 cup of sugar.

Butter amounts varied from 2 tablespoons to 8 tablespoons. I found that the amount of butter resulted in much more subtle differences than the amount of sugar. Curd made with only a little butter was more tart and fruity. The more butter I added, the more complex the flavors became, but also the more the fruit flavor was masked. I preferred about 6 tablespoons of butter. I can certainly see the argument in favor of less, though, particularly if you want a cleaner, fresher flavor.

I also made curds with different citrus fruits. Although lemon was the exemplar (and Meyer lemon even better), lime was very nice, with a pointed acidity and a slightly grassy, herbal quality. Orange was softly acidic and delicately floral. I tried blood oranges, too, but that was the only washout. When blood oranges are cooked, the vivid red color fades to a bruised purple, and the intriguing berry quality becomes nasty and artificial-tasting.

Most interestingly (at least to a food wonk), I tried to make curd with plain water to test the effect of liquid with no acidity and wound up with sweet, watery scrambled eggs. Acidity lowers the coagulation temperature and obviously plays a very important role in tempering the action of the sugar.

The most amazing thing about curd isn't its many variations but the one thing all versions have in common: curd is dead simple to make. In fact, as delicate as it may seem, you don't even need to use a double boiler; you can make it in a single small saucepan. Just use a common pastry chef's trick: bring the mixture quickly to a boil over high heat, then strain out any bits of curdled egg. I recommend a path somewhere between the cautious double boiler and the bold chef's methods — cooking the curd over medium heat while using cold butter to moderate its temperature, then reducing the heat once the butter melts and continuing to cook until the curd thickens.

When the curd is done, it should coat the back of a spoon like a moderately thick hollandaise. It should be thick enough that when you draw your finger across the spoon, you divide the coating into two distinct sections. It will set more as it cools. Curious about the temperature, I measured it a couple of times and found that the mixture began to smooth out and thicken slightly at about 165 degrees and was finished at about 185 degrees.

The key to success is making sure the eggs and sugar are well beaten before you add the remaining ingredients. That's the only way to temper the proteins sufficiently. When I tried making curd by simply mixing everything together, I got fine bits of curdled whites, even after straining.

Meyer Lemon Granita

If you've ever wondered what's so special about the Meyer lemon, taste this granita, which perfectly captures the fruit's character. If you can't find Meyer lemons, add 3 tablespoons mandarin (tangerine) juice to 5 tablespoons regular lemon juice.

6 SERVINGS

1½ cups water
1½ cups sugar
 2 tablespoons grated Meyer lemon zest
    (about 5 lemons)
 1 cup Meyer lemon juice

Heat the water and sugar just until clear, about 5 minutes. Add the lemon zest and remove from the heat. Let the syrup steep for at least 30 minutes.

Stir in the lemon juice and strain the mixture into a shallow metal dish, such as a cake pan. Freeze for 1 hour, then stir the mixture with a fork, breaking up any chunks of ice. Repeat 4 or 5 times over the next 2 to 3 hours. Each time the ice will be a little less liquid and will stick together more. When it is firm enough to hold a shape, it is done.

Try not to let the ice freeze solid. If it does, chop it into small pieces in the pan and grind it in the food processor. The result will be lighter and fluffier, and the flavor will not be as intense.

Mandarins (Tangerines), Grapefruits and Pummelos

· ·

It has long been a fruit seller's dream: a sweet piece of citrus that has no seeds to spit out, that peels as easily as a candy bar and that is available for at least half the year. Could there be a more convenient snack food? Ironically, the newest thing in citrus is also one of the oldest.

Mandarins (we used to call them tangerines) are one of the three original citrus families — the other two being the seldom seen citrons and pummelos. All the other citrus — including oranges, lemons, limes and grapefruits — are hybrid results of crosses between those three groups.

As you might expect with a heritage so long, the family tree of the mandarin is incredibly diverse. One leading citrus botanist divides mandarins into thirty-five separate groups (and each group into dozens, even scores, of separate varieties). The signal feature of all mandarins is a thin peel that is easy to remove, rather than clinging tightly to the inner fruit as is true with other citrus. This trait is variable: the skin of some mandarins is so loose that at maturity it touches the fruit at only a couple of spots; other mandarins are only marginally easier to peel than an everyday orange. The Clementine — the original "tangerine," so called because the first fruit was imported to the United States via the Moroccan port of Tangier — has been around for more than a century and can be seedless. The Satsuma, a Japanese mandarin, is even older. Most Satsuma varieties are reliably seedless.

Shirttail cousins include crosses between pure mandarins and other citrus. The most popular are tangelos (Minneola being the prime variety), which are the result of crossing mandarins and pummelos, which look like giant grapefruits. Tangelos are larger than most mandarins but also are easy to peel. They usually have a small "neck" at the stem end. The tangor, or Temple orange, is probably the result of a cross between a mandarin and an orange, although some experts hold out the possibility that it is a variety of tangelo.

The hard part has been getting a seedless mandarin to market later than mid-January. Until recently, the vast majority of American mandarins came from Florida and with the exception of the Murcott (popularly called the Honey), mandarins were tough to find later than Christmas. Even the Murcotts were almost always done by early March. There were other complications, too. Clementines, for example, are seedless only when they are grown in isolation, away from other citrus. This is extremely inconvenient for growers, who like to have a range of fruits to offer.

Until the early 1990s, mandarins represented a minor part of the American citrus market — no more than 5 percent a year. Then came a wake-up call from Spain. The largest exporter of fresh citrus in the world, Spain started sending boatloads of Clementines to the United States and found a downright eager market. From 1996 to 2000, Spanish shipments of Clementines increased fivefold, to more than 200 million pounds a year.

That got the attention of California's citrus growers, many of whom were getting squeezed in the souring orange market. California has the advantage of a cooler growing climate than Florida, so it can produce mandarins much later in the year. California growers began planting great swaths of mandarins in the southern part of the San Joaquin Valley in the late 1990s, more than doubling the state's previous plantings in just five years and drawing nearly even with Florida. From 1998 to 2005, California's acreage increased from about 8,000 acres to more than 18,000, and Florida's plantings decreased. During the same period plant breeders at the University of California at Riverside released a series of new seedless mandarin hybrids that have the potential to extend the season well into early summer.

Although Spain produces about 2 million tons of mandarins every year and still dominates the global market, the United States grows 40 percent of the grapefruits in the world. (The name comes from the fact that the decidedly ungrapelike fruit grows in clusters at the ends of branches.) The origin of the grapefruit is unclear, but most botanists believe it is a cross between the pummelo and a sweet orange. It was first described on a plantation in Barbados in 1837 and was introduced in Florida not long after.

Grapefruits are beloved by growers and retailers for their remarkable durability. The ripe fruit of most varieties will hold on the tree for months, waiting for the most opportune time for harvest. Once picked, grapefruits will last for weeks, even when kept at cool room temperature.

Florida grows more than 80 percent of the U.S. harvest, primarily the white Marsh variety. Texas is a distant second, with less than 15 percent of the Florida total. The state is best known for its Star Ruby and Ruby Red deep pink grapefruits, which get their color from the flavorless pigment lycopene. These grapefruits turned up in the early twentieth century as chance mutations and quickly found ready consumers.

The pummelo, a related fruit, can be as big as a human head and has thick skin with lots of coarse white pith. Most pummelos taste very sweet, primarily because they are so low in acid. The fruit is extremely dry, so much so that the most common way to prepare it is by breaking each segment into individual vesicles — those little beads of fruit that are practically invisible in richer, moister citrus.

A delicious grapefruit-pummelo cross in California called the "Oroblanco" is slowly finding an audience, thanks to a particularly circuitous marketing path. Originally introduced in the early 1980s, the Oroblanco was praised for its fine flavor — it is very sweet but is balanced by a bracing astringency. Because the fruit is slow to color, it fell from favor after only a few years. Then Israeli growers renamed the fruit "Sweetie" and began exporting it to Japan, basing an ad campaign on the fact that this fruit is sweet even when green. Now California growers are beginning to give the variety another chance.

WHERE THEY'RE GROWN: Mandarins are split almost equally between Florida and California, with a significant number of Clementines being imported from Spain during the holidays. Florida dominates the grapefruit market in terms of sheer numbers, but there are also important plantings in Texas, California and Arizona.

HOW TO CHOOSE: The skins of mandarins should feel firm, not wrinkled. This is a little harder to discern with some varieties, which "puff up" when ripe, with the skin separating from the flesh, but there should certainly be no soft spots. The fruit should be heavy for its size, and the fragrance should be clean and fresh. Mandarins that are past their prime will smell a little fermented. The skins of grapefruits should be smooth, and the fruit should be very heavy. Check the stem ends as well — when the fruit gets a little old, the end will start to sink into the fruit.

HOW TO STORE: Because of their thin skins, mandarins should be refrigerated as soon as you get them home. Do not wash them, as water will hasten the breakdown of their peels. Grapefruit have thicker skins and so will last a little longer at room temperature. This is particularly true for varieties with a lot of pummelo heritage — their rinds can be so thick that they will last for weeks.

HOW TO PREPARE: Mandarins usually are easy to peel, but there may be a lot of stringy white pith that remains attached to the fruit. Since it will fall off and look messy in a dish, it should be removed as well. Simply run your thumbnail along the outside of the fruit; the strings will pull away. Have a damp towel nearby to wipe the sticky strings from your fingers. Peel grapefruits with a sharp knife, just as you would oranges. The varieties that have thicker skins are even easier to peel: cut off the top and bottom, score the skin and pull the peel away with your thumbs.

ONE SIMPLE DISH: After a big meal, I love to serve a bowl of mandarins with dried dates or prunes or with simple sugar cookies. There are few smells more satisfying than the fragrance when you peel them. Serve grapefruits the same way, but peel them first.

Mandarins with Rosemary Honey

The combination of rosemary and mandarins may seem a little odd, but it tastes and looks fantastic. This dish is even more attractive if some of the blue flowers are still attached to the rosemary sprigs.

6 SERVINGS

- ½ cup wildflower honey
- 1¼ teaspoons minced fresh rosemary leaves
- 1½ pounds seedless mandarins (about 9)
  Rosemary sprigs for garnish

Heat the honey and minced rosemary in a small saucepan over low heat until the honey is liquid. Set aside to steep for at least 15 minutes.

Peel the mandarins and break them into bite-size sections of roughly 2 segments each. Carefully remove as much of the white pith and string as you can. Notice that there is a central string that runs down the back of almost every segment. Remove this, and much of the pith will come with it.

Arrange the mandarins in a low mound on a serving platter and spoon the rosemary honey over the top. Each piece of mandarin should be touched by, though not coated with, the honey. Garnish with the rosemary sprigs and serve.

Mandarin Parfait with Candied Ginger

I especially like this recipe with tiny Clementines or Satsumas, which are exactly the right size.

6 SERVINGS

- 2¾ cups whole milk
- 1 large egg
- ¼ cup sugar
- 3 tablespoons instant tapioca
- ¼ teaspoon salt
- 1 vanilla bean
- 1 cup sliced almonds
- 1 pound seedless mandarins (about 6)
- 2 tablespoons minced candied ginger (about 1 ounce)

Whisk together the milk, egg, sugar, tapioca and salt in a saucepan over medium heat until the egg is well combined. Remove from the heat and set aside for 5 minutes. Then bring to a rolling boil over medium heat, stirring frequently. Split the vanilla bean and scrape the seeds into the tapioca mixture. Remove the tapioca from the heat, pour it into a container, cover and refrigerate until set, at least 1 hour. The mixture will be somewhat liquid at first, but it will thicken as it cools.

Heat the oven to 350 degrees. Spread the almonds on an ungreased jelly-roll pan and toast, stirring occasionally, until they begin to become fragrant, 7 to 10 minutes. Shake the pan occasionally to prevent scorching. When the almonds are fragrant and beginning to turn golden, remove them from the oven and immediately transfer them to a bowl to stop the cooking. Set aside.

Peel the mandarins and carefully remove as much of the white pith and strings from the fruit as you can. Break the mandarins into sections containing roughly 2 segments each and cut each section in half crosswise. Collect the mandarins in a bowl and set aside.

Sprinkle the minced ginger over the chilled tapioca and stir roughly to mix it in thoroughly and to break up the tapioca. After mixing, the tapioca should be somewhat creamy in texture rather than set firm like Jell-O.

Spoon 2 to 3 tablespoons of tapioca into each of six small glasses. Martini glasses are perfect; wineglasses or flat champagne coupes work well, too. Spoon half a dozen pieces of mandarin over the top, then scatter a generous tablespoon of sliced almonds over that. Repeat twice more so that you have 3 succeeding layers of tapioca, mandarins and almonds. Serve cold or refrigerate for up to 1 hour before serving.

Oranges

. . . .

One of our most common fruits has a most unusual history. The path that led to the development of the modern orange was a long and circuitous one, with many curious twists and turns. In the beginning, there was no such thing as an orange. Sweet oranges are believed to be a cross between the pummelo and the mandarin. The first cultivation of the sweet orange was in China, where it thrived for several centuries before spice traders from the West brought it to Europe via Genoa in the early 1400s. It was quickly accepted, particularly by the Portuguese, who were at the height of their commercial and naval powers. Their efforts introduced the fruit around the Mediterranean basin, and for centuries one of the most popular nicknames for the sweet orange was "Portugal orange." By the early sixteenth century, the orange was well established, being grown in the eastern and southern Mediterranean as well as in Portugal, Spain and Italy. Oranges were highly valued, and new growing areas were eagerly sought. Columbus almost certainly planted oranges in what is now Haiti on his second voyage to the New World, in 1493. The fruit probably reached Central America not long after; one reference mentions orange trees being planted in Mexico in 1518. Thenceforth, wherever conquistadors and missionaries traveled, the orange followed, including California and, most notably, Florida.

With the waning of Spanish influence in North America in the eighteenth century, citrus growing declined in Florida. In the mid-1800s, when a new generation of farmers came to Florida after the territory was acquired by the United States, large groves of wild orange trees were all that remained of the extensive orchards planted by the Spanish. Being enterprising sorts, the newcomers simply grafted modern varieties onto the existing trunks and started again. By the end of the nineteenth century, Florida was once again the king of citrus in America, producing almost four times the amount of its nearest competitor, California.

Two unrelated but roughly coincident events turned the citrus world on its head. The first was the introduction of a new type of orange, the Washington navel, in California in the mid-1870s. The second was a natural calamity in Florida. In the winter of 1894, a ferocious cold snap hit the citrus orchards there, freezing thousands of orange trees down to the ground. It took more than a decade for the Florida harvest to return to its prefreeze levels. In the meantime, the navel orange took hold in California.

Originating from a chance mutation in the Bahia region of Brazil, probably in the early nineteenth century, the Washington was almost immediately recognized as one of the world's best-tasting oranges, despite its temperamental growing habits and its relative lack of juice. (In contrast, the Valencia, the dominant fruit of Florida, is extremely juicy, composed of almost 50 percent juice, but the fruit's flavor is relatively undistinguished.) The navel quickly spread around the world, being adopted in Paraguay, Spain, South Africa, Australia and Japan.

Navels have been known for centuries in the Mediterranean. They are typified by a "bellybutton" that appears at the flower end of the mature fruit. This is actually a small, primitive secondary fruit embedded in the larger orange. Besides their flavor, navel oranges have several qualities to recommend them: they are frequently seedless, they are relatively easy to peel and they have flesh that is firm rather than watery. When planted in Florida, however, navels become coarse and granular. Furthermore, when the fruit is juiced, a compound called limonin turns it bitter within half an hour. Increasingly, Florida growers concentrated on producing juice from their Valencias and similar varieties, and California growers concentrated more on fresh fruit. Today almost 97 percent of Florida's oranges wind up in juice, and chances are good that you have never eaten a Florida orange.

Florida's change from fresh to processed did not occur all at once, but over decades. The clincher came in the late 1940s and early 1950s, when new processing technology allowed the creation of frozen juice concentrate, which tasted better than canned juice. To make frozen concentrate, the fruit is squeezed and the juice is heated to destroy enzymes that might spoil its flavor. Then the juice is put in a vacuum extractor that evaporates out most of the water. The concentrated juice (now containing roughly four times as much sugar as fresh squeezed) is then frozen until needed. Finally, fresh juice is mixed back in with the frozen, sometimes with extracted orange oil from the peel to reinforce the flavor.

In addition to Valencias and navels, the family of oranges includes other juicing varieties; blood oranges, which have ruby-colored flesh and a raspberry-like flavor; and the highly popular eating orange called the Cara Cara. A chance mutation found in 1976 on a Washington navel in a Venezuelan orchard, the Cara Cara is a sweet, low-acid orange whose flesh is a deep pinkish orange — close to the color of Ruby Red grapefruit.

WHERE THEY'RE GROWN: Most fresh oranges for eating are grown in California. The main growing area was once in the southern part of the state, particularly around the town of Riverside and in Los Angeles and Orange counties. Today expanding urban centers have pushed the orange groves north to the Central Valley, where they are usually located on the western slope of the Sierra. There they catch a nice evening chill that improves their color and flavor. A small number of fresh oranges are also grown in Arizona.

HOW TO CHOOSE: Choose oranges that are heaviest for their size. In most cases, pay no attention to the color of the skin, which can be influenced by variety and also by dyeing at the packing shed. Sometimes, in fact, perfectly ripe oranges will have green patches on the skin. This usually happens when the fruit hasn't been picked before the tree begins to blossom for the next season — something that is not uncommon with California Valencias. This process is called regreening and does not affect the quality of the fruit.

HOW TO STORE: Because oranges have relatively thick skins, they can be stored at room temperature for up to a couple of weeks. If you buy them in bags, take them out so the trapped moisture doesn't lead to mold. Refrigerating oranges does not hurt the flavor.

HOW TO PREPARE: You can, of course, peel an orange in the time-honored way, but if you're cooking, it's neater to use a knife. Cut an even slice from the top and bottom and set the orange on a cutting board. Cut one vertical strip from the orange, beginning at the point where you see the white pith meet the orange flesh and following the natural curve of the orange. Repeat, working your way around the orange. With each slice, use the newly exposed vertical line between pith and flesh to guide you. When all of the skin has been removed, carefully slice away any bits of white pith that remain.

ONE SIMPLE DISH: Make a vanilla-scented syrup by boiling 1½ cups water and 1 cup sugar with a split vanilla bean until the syrup is clear, about 10 minutes. Refrigerate. Peel the oranges and slice into a bowl, then pour the cold syrup over the top. Remove the vanilla bean. (You can stick this in your sugar canister to make vanilla sugar.) Serve with crisp cookies.

Orange and Beet Salad
with Goat Cheese and Walnuts

This recipe is as open-ended as a jazz score. Try using golden beets instead of red ones. Roast the beets or steam them. Use blood oranges instead of, or in addition to, navels, or mix in some grapefruit sections. Instead of the goat cheese, you can substitute ricotta salata. The dish is also nice with a few black olives tossed in.

To roast the beets, cut them back to about an inch of the stems, wrap them in aluminum foil and bake them at 400 degrees until they are tender when pierced with a knife, about 50 minutes. Alternatively, you can steam the beets, which will take only about 25 minutes. Once the beets are cooked, the skins will slip right off.

6 SERVINGS

- 1 pound cooked beets (see headnote)
  Pinch powdered oregano
  Salt and freshly ground pepper
- 2 tablespoons olive oil, plus more for drizzling
- 1 teaspoon sherry vinegar
- ¾ cup walnut halves
- 6 navel oranges
- 1 ½-pound log fresh goat cheese
- 2 tablespoons chopped fresh parsley leaves

Heat the oven to 450 degrees. Slice the beets about ¼ inch thick and place them in a bowl. Season with the oregano and with salt and pepper to taste. Add the olive oil and sherry vinegar and stir to mix well. Set aside.

Place the walnut halves in a dry baking pan and toast in the oven, stirring occasionally, until they smell toasty and nutty, about 5 minutes. Chop coarsely, cutting each half into 3 or 4 pieces. Set aside.

Peel the oranges. Cut off the top and bottom to expose the flesh. With a thin, sharp knife, slice away a vertical section of the skin, following the shape of the orange. Make a second cut, following the line of exposed flesh created by the first. Repeat all the way around the orange until all of the skin has been removed. Carefully go back over the orange, cutting away any small pieces of pith that remain. Slice the oranges ¼ inch thick and arrange them in a low mound on a platter.

Spoon the beets in a low mound in the center of the oranges, leaving the oranges exposed along the edge.

Slice the goat cheese into thin disks (as thinly as you can and still have them retain their shape). Scatter the disks over the beets. Season with a little more pepper and lightly drizzle with olive oil. Sprinkle the toasted walnuts and parsley on top and serve.

Old-Fashioned Orange Cake

This "momma cake" is moist in texture, rich with chunks of oranges, raisins and walnuts and fragrantly spiced with cinnamon. I learned it from Marion Wilke, who entered it in a citrus cooking contest I judged.

10 SERVINGS

1 navel orange
1 cup raisins
1 cup sugar
½ cup shortening
2 large eggs
¾ cup plus 2 tablespoons buttermilk
2 cups cake flour
1 teaspoon ground cinnamon
1 teaspoon baking soda
½ teaspoon baking powder
1 cup chopped walnuts
1 cup confectioners' sugar

Heat the oven to 325 degrees. Grease a 9-by-13-inch baking pan. Scrub the rind of the orange well. Cut off both ends of the orange, then cut it into large chunks. Set aside.

Place the raisins in a food processor and pulse 6 or 7 times until coarsely chopped. Add the orange chunks and pulse 5 or 6 times to reduce them to equal-size small pieces. Empty into a small bowl and set aside.

Pulse the sugar and shortening together in the food processor 4 or 5 times to combine, scraping down the sides of the bowl. Add the eggs one at a time and process in between, scraping down the sides. The mixture should be smooth, light and fluffy. Add ¾ cup of the buttermilk and process to combine. The mixture will be very liquid and may look slightly separated.

Sift together the flour, cinnamon, baking soda and baking powder and add to the processor. Pulse just to combine into a smooth batter. Set aside 2 tablespoons of the orange-raisin mixture for the topping. Add the remaining orange-raisin mixture to the processor and pulse once or twice. Add the walnuts and pulse once; do not overprocess.

Pour the batter into the prepared pan. Bake until the sides have begun to brown and pull away from the pan, 35 to 40 minutes. The cake should still be a little moist in the center.

Whisk together the remaining 2 tablespoons buttermilk and the confectioners' sugar until smooth. Whisk in the reserved orange-raisin mixture and spread the icing over the hot cake. Serve at room temperature.

Candied Citrus Peel

I make only one kind of candy: candied citrus peel. The reason is simple: I've never tasted a version that held a candle to homemade. The following method, developed from a composite of recipes, gives the best flavor balance with a minimum of effort.

5 **pounds citrus fruit (oranges, grapefruits and/or pummelos)**

5 **cups sugar**

4 **cups water**

Score the skin of each citrus fruit in sections about 2 inches wide, cutting through the skin but not into the fruit. Use your fingers to peel the fruit, carefully running your thumb between the skin and the fruit to separate them. The pith will remain attached to the skin. Reserve the fruit for another use.

Put all the citrus peels in a large pot and cover with water. Bring to a boil, drain the peels and rinse briefly under cold water. Blanch the peels in the same manner 2 more times. After the third blanching, drain the peels and rinse them under cold water. Using a thin, sharp knife, remove as much pith as possible. You should be able to see some color through any remaining pith. Cut the peels into shreds ⅛ to ¼ inch wide.

While the peels are blanching, bring 4 cups of the sugar and the water to a boil in a large saucepan. Cook, stirring occasionally, over medium heat until you have a thin syrup, about 1 hour. You will have about 4 cups syrup.

Cover the shredded citrus peel with cold water and bring to a simmer. Cook until the peel begins to lose its raw look, 5 to 10 minutes. Drain and immediately, without rinsing the peel, transfer it to a large bowl. Cover the hot peel with the hot syrup and set aside for 1 hour to candy.

Heat the oven to warm. After the fruit has candied, place the remaining 1 cup sugar in the bottom of another large bowl. Drain the peel, add it to the sugar and toss to coat well. Shaking to remove any excess sugar, transfer the peel to a wire cake rack set over a jelly-roll pan to catch the sugar. Arrange in as thin a layer as possible. Place the rack and the pan in the oven for 30 minutes to dry.

Remove the peel from the oven and let it sit at room temperature to finish drying. The peel will become firmer and chewier over several days. When the peel is just the way you like it, place it in an airtight container. Store at room temperature for up to several months.

Index

H